Data Analysis with Machine Learning for Psychologists

Chandril Ghosh

Data Analysis with Machine Learning for Psychologists

Crash Course to Learn Python 3 and Machine Learning in 10 hours

Springer

Chandril Ghosh
Department of Psychology
Bath Spa University
Newton Park Campus, Bath, UK

ISBN 978-3-031-14633-6 ISBN 978-3-031-14634-3 (eBook)
https://doi.org/10.1007/978-3-031-14634-3

This Springer imprint is published by the registered company Springer Nature Switzerland AG
The registered company address is: Gewerbestrasse 11, 6330 Cham, Switzerland

This book is dedicated to the Psychologists of the future.

Acknowledgement

This book is a collection of ideas from the fascinating field of machine learning. I am thankful to all those researchers and practitioners of machine learning who conducted the research that I have picked and introduced in this book. Without their work, this book would not had been possible. There are also several anonymous authors on websites like Stack Overflow whom I could not cite but their advice, ideas and codes continue to inspire me and this book. I must acknowledge their contribution too. I had also been fortunate to receive insights and feedback from my colleagues, and students. In particular, I would like to thank Mr Richard Joltes, Mr Varun Thakre and Prof Deepak Padmanabhan for reviewing and providing feedback on various chapters of the book. I should thank my parents Chandan and Papiya Ghosh, and my partner Madhumanti Das for their support and faith in me. Finally, I must value and appreciate the constant support of the Project Coordinator Ms. Olivia Ramya Chitranjan, (ex) Senior Editor Lilith Dorko and the team of Springer. They have helped me make this book a reality.

Contents

1 Introduction 1
1.1 Before Getting Started 3
1.1.1 Overview (of Old Ways to Analyse Data and Some Problems Related to Them) 3
1.1.2 Who Am I? 6
1.1.3 How Did I Get There? 6
1.1.4 Who Is This Book For? 7
1.1.5 Who Is This Book NOT For? 8
1.1.6 What's Special You Get in This Book? 9
1.1.7 So, What Does This Book Have? 9
1.1.8 How to Best Make Use of This Book? 9
1.2 Types of Research Studies 10
1.2.1 Explanatory Research 11
1.2.2 Predictive Research 15
1.2.3 Exploratory Research 17
1.3 Data 17
1.3.1 To Collect or Not Collect Your Own Data 17
1.3.2 Where to Get the Data From? 19
1.3.3 Ways in Which Data Is Divided 23
1.3.4 Five Lessons 24
1.4 Statistics: A Refresher Before Getting into Machine Learning 26
References 30

2 Python Programming 33
2.1 But Do I Have to Learn to Code for Data Analysis? 34
2.2 How to Install Python? 34
2.3 Variables 38
2.4 Operators 41
2.4.1 Arithmetic Operators 41
2.4.2 Comparison Operators 43

2.5 Statements ... 44
2.6 Loops ... 44
2.7 Data Structure ... 47
2.8 Methods and Functions (Built-Ins) in Python ... 48
2.8.1 Methods ... 48
2.8.2 Function ... 50
2.9 Error Resolution ... 51
2.10 Last Words ... 52

3 Data Pre-processing ... 55
3.1 Introduction ... 55
3.2 Data Cleaning ... 58
3.2.1 Problem 1: Duplicate Columns and Categorical Variables ... 59
3.2.2 Problem 2: Outliers ... 62
3.2.3 Problem 3: Missing Values ... 64
3.3 Data Transformation ... 67
3.3.1 Converting Categorical Variables into Numeric Variables ... 67
3.3.2 Converting Continuous Variables into Categorical Variables ... 67
3.3.3 Feature Scaling ... 70
3.4 Data Reduction ... 74
3.4.1 Strategy 1 ... 77
3.4.2 Strategy 2 ... 78
3.4.3 Strategy 3 ... 78
3.4.4 Strategy 4 ... 79
3.4.5 Strategy 5 ... 82
3.5 Final Words ... 84
References ... 85

4 Machine Learning ... 87
4.1 Introduction ... 88
4.2 Classification ... 101
4.2.1 Getting Started with Supervised Machine Learning ... 101
4.2.2 Machine Learning (Classifier): The Leak-Proof Approach ... 112
4.2.3 Confidence Interval ... 116
4.2.4 Choosing the Best Model for Classification ... 119
4.2.5 Optimising the Predictive Accuracies of the Model with Hyperparameter Tuning ... 123
4.3 Regression ... 127
4.3.1 Regression Using Machine Learning and How to Interpret the Results ... 127
4.3.2 Feature Importance ... 130

4.3.3 Exploratory Research Using Unsupervised Machine Learning 136
4.4 Clustering 136
4.4.1 Hierarchical Clustering 136
4.4.2 K-Means Clustering 143
4.5 Principal Component Analysis (PCA) 148
4.6 Rule Mining 149
References 154

5 End Note 157

Index 161

About the Author

Chandril Ghosh is a UK-based chartered psychologist and is currently working as the Lecturer in Clinical/Counselling Psychology at the Bath Spa University. He completed his BSc in psychology (with honours) and MSc in Clinical Psychology from India. After completing his MSc, Ghosh began to study machine learning and Python programming through books and online materials on the subject. He had no background or prior experience with coding or computer science back then. During his doctoral studies, he utilised his knowledge on the subject to employ machine learning techniques to explore psychopathology. You will find his work published by the BMC Psychiatry journal. Around the same time, he was hired multiple times to design and deliver a crash course on Python 3 and machine learning for post-graduate students at Queen's University Belfast. Furthermore, he continues to run online courses on the subject outside the University and teaches students from about 56 countries. This book is a product of an accumulation of his hundreds of hours of teaching and feedback from students with social science backgrounds.

Chapter 1
Introduction

Preview

I know you are here to learn how you can implement machine learning in your research as quick as possible. So, to cut the chase and keep it short, here's a sample of a machine learning code script written in Python (Fig. 1.1). You can use this code on a dataset of patients with individual symptoms. When you use this script on the data, it will generate a plot that will tell you which people have more in common with each other than others. So, for example, if you have ten patients with 12 symptoms (quantitative survey responses), this code will tell you which of these patients share similar symptoms and which pairs of people are dissimilar in their experience. I imagine this to find its place in a study investigating diagnostics in mental healthcare.

Take a close look at the script in Fig. 1.1. There are five main steps:

- Step 1: Import libraries (not your school library!)
- Step 2: Load the dataset
- Step 3: Normalise the data
- Step 4: Plot the dendrogram
- Step 5: Create agglomerative clustering

It is totally fine if you don't understand any of these. That's natural – otherwise, why would you be reading this book now?

Being bombarded with the code script first thing in this book, you might be haunted by the Python codes on this script. So, we will learn Python codes first. That will enable you to understand what each line of this code script means. It's simple. Trust me. You will also understand these libraries and why we are importing them.

Secondly, you might notice that in step 3, there's something called normalising the data. Assuming that you have learnt statistics, you probably know a bit of what it is, but if not, don't worry; you will learn it here. But for now, understand that raw data (one that we get from the participants) has a trash-like quality in itself. It's like

C. Ghosh, *Data Analysis with Machine Learning for Psychologists*,
https://doi.org/10.1007/978-3-031-14634-3_1

```
In [22]: #Step 1: Import libraries
import pandas as pd
import scipy.cluster.hierarchy as cluster
import matplotlib.pyplot as plt
from sklearn.preprocessing import normalize
from sklearn.cluster import AgglomerativeClustering

#Step 2: Load the dataset
data = pd.read_csv("C:\\Users\\ghosh\\Downloads\\Pre-processed dataset.csv", header=0, index_col=0)
variable_names = ["Not being able to stop or control worrying", "Little interest or pleasure in doing things", "Feeling down, de
X = data[variable_names]

#Step 3: Normalise the data
scaler = MinMaxScaler()
normalised_data = pd.DataFrame(scaler.fit_transform(X))
normalised_data.columns = ["Not being able to stop or control worrying", "Little interest or pleasure in doing things", "Feeling
normalised_data.describe()


#Step 4: Plot the dendrogram
plt.figure(figsize=(10, 7))
plt.title("Dendrogram Demo")
dend = cluster.dendrogram(cluster.linkage(normalised_data, method='ward'), leaf_rotation=90)

#Step 5: Create an instance of Agglomerative-Clustering
modelx = AgglomerativeClustering(affinity='euclidean', linkage='ward')
modelx.fit(normalised_data)

#Step 6: Return an array of integers corresponding to each categories or clusters
#print (modelx.labels_)

#plt.savefig('plt.png', dpi=520, format='png', bbox_inches='tight')
```

Fig. 1.1 Code for hierarchical clustering

crude oil. You can't run your car on crude oil. So, we refine that crude oil, turn it into petrol and can pour that into our car. Likewise, raw data needs to be refined, and so we will learn how to do data pre-processing. Spoiler alert – Normalisation is one way to clean the data.

Finally, it's time for us to implement the machine learning model and generate plots, as you saw in steps 4 and 5. You would love to publish these plots in our research article's "result" section. We use hierarchical clustering (agglomerative sub-type) in the code you just saw. There is a dozen of other machine learning models for different purposes. Yes, you will learn them all (or at least the popular ones).

By the end of this book, you will be able to perfectly understand code scripts like the one in Fig. 1.1 and learn a dozen others. But as you might have guessed by now, we need to take three steps – one by one – to reach there.

- Step 1: Learn how to code in Python (Chap. 2)
- Step 2: Learn how to clean the data using data pre-processing (Chap. 3)
- Step 3: Learn how to implement machine learning models for your own projects (Chap. 4)

The first chapter consists of the customary introduction, why I (as an author) am a reliable source, who this book is for, how to make the most out of this book, some refreshers about statistics, research methodology and blah blah.

If you are sold on my idea for creating this book and have a busy schedule to follow up, then jump straight to the "meat" on Chap. 2 and get started with Python programming. Otherwise, stick around for a while through the first few pages to get a little deeper insight into the background of this topic.

1.1 Before Getting Started

1.1.1 Overview (of Old Ways to Analyse Data and Some Problems Related to Them)

For generations, psychologists have been trained in traditional numeric statistics to describe the sample and infer population reports/opinions from sample data. But with the emerging new data sources and challenges psychologists face in producing application-oriented research outputs, it is now the time to expand the repertoire of the range of tools that can be used to analyse the dataset.

- Since much of our psychological literature is based on surveys and interviews or observation, the size of data was mostly small to moderate. I remember how happy we got after interviewing 26 participants or after surveying 400 people. That was the best we could do until now. But with the advent of technology, such as smart watches, social media and mobile phone apps, alternative sources of massive amounts of data are getting increasingly available to psychologists to study human behaviour. Conventional ways of analysing data we learnt in graduate schools may not be sufficient in the era of big data. So, what sort of problems?
- **Format of data**: Statistical investigations (e.g. multiple regression) are fit for numeric data values. However, big data collected from alternative sources such as the Internet often have different data formats such as numbers, texts, sounds, images and videos. Conventional multivariate techniques might not be able to handle those.
- **Conclusions of the analysis**: Often, psychologists use the significance test to judge the relevance of the predictor variables (for example). But as the sample size grows large, it may lead to statistically significant results in most instances (even when the associated effect sizes are very low).
- **Reproducibility**: Some of the current literature also raises concerns about the replication and reproducibility of psychological research analysed using statistical inference (Orrù et al., 2020).

As you might have already guessed, machine learning can successfully address all the abovementioned issues (Cheung & Jak, 2018; Orrù et al., 2020). That said, I do not mean machine learning is always superior to traditional numeric statistics. Instead, both of these are better suitable for different kinds of problems and datasets. Take the following two as examples:

1. Traditional numeric statistics primarily focus on the parameter estimates (i.e. coefficients of the predictor variables). But machine learning primarily focuses on the accuracy of prediction. There are times when traditional numeric statistics also focus on accuracy. But machine learning can aggregate large quantities of data in less time and is also less subjective. So, depending on your goal, you can pick a suitable approach. That said, such distinctions are good for heuristics. Still, they are not rules written in stone because you *can* focus on the predictors

while using machine learning techniques. You will learn to figure out which are the most important predictors of a target variable later in the feature importance lesson of this book. But for the sake of heuristics, we can argue that traditional numeric statistics is concerned about the input and machine learning is concerned about the output.

2. Another way to differentiate these two different data analysis techniques was that machine learning techniques performed better than multiple regression when the data structure was complex with nonlinear patterns, but multiple regression worked well when choosing predictors less relevant to the criteria. Also, you will sometimes find those statistical techniques in the machine learning toolbox. So, the difference between traditional numeric statistics and machine learning is often fuzzy. Machine learning uses statistics, just in a different way. It's evaluating a dataset probabilistically, which has been argued to be a just another spin on statistical analysis. In fact, time and again, I have come across several researchers using both statistics and machine learning to address different questions within the same study.

Now is the time to give a reason for what's in it for you to read this book thoroughly and learn machine learning using Python today. From the above discussion, I hope you can imagine that there are kinds of data or research questions that are not answered by statistics. Therefore, reading this book till the end and practising all the exercises will enable you to investigate new lines of enquiry. That's the first reason. But that's not all. There are more reasons for you to complete this book on machine learning for data analysis.

The second reason is that learning machine learning (and Python programming) makes you more employable and is considered a desired skill (Whiting, 2020). Within academia, more and more job recruiters are asking for machine learning skills in the universities' psychology departments. Likewise, neuroscience and experimental psychology departments often look for Python programming skills to develop psychology experiments on the computer in their new recruits. Consider familiarising yourself with PsychoPy, an open-source Python-based software meant to develop experiments outside this book. Even outside academia, your scope of getting a job increases. For example, many psychology-based start-ups are popping up in the world today (e.g. mental health apps). Otherwise, there are always openings for psychology students to apply for research or business/marketing/HR analytics roles in businesses, even if you don't have a formal background in HR or marketing or any other aspect of a business, but still have training on how to understand and analyse human behaviour – which is what businesses want at the end of the day to attract new customers or to retain the best talents (for example). So, if you follow through with this book and practise the exercises by the end of it, you will get a bit more employable than you are right now.

The third reason is that if you are in academia, then know that academic papers with "artificial intelligence" get more citations than any other research topic (Crew, 2019). Of course, the quality of the study, the journal and the field of your work are essential variables, but let's say you can expect to get higher citations than you are getting now. Plus, grant committees and governments are increasing their funding in this domain (European Commission, 2019).

To sum up, completing this book increases your chances of getting employment, improves your chances of getting a grant approval and enables you to publish papers that get higher than average citations.

There are many ways to teach Python for machine learning. In this book, my primary focus will be teaching how to read Python scripts and modify different elements to fit your use case. There are proponents of the idea that you should learn to write Python codes from scratch. I really like that idea, but I find it extremely demanding for the students. More importantly, it is unnecessary most of the time. After all, we are not computer scientists or software developers but psychologists who use these codes for their academic pursuits. So, what we are required to know to be able to analyse the datasets using Python and machine learning is also very limited compared to the industrial use cases a Python developer might need to face or is expected to do. Keeping that in mind, I have tried to create a shortcut path for you to learn how to quickly implement machine learning for data analysis. There are Python codes for almost every possible kind of data analysis available on the Internet for free of cost. Suppose you know how to read those codes. In that case, you can simply copy them from the Internet pages and paste them on your computer, then modify them appropriately to run on your own data. While this will work for you most of the time, to err on the side of caution, I must mention that while this is true, it can be hazardous to treat this as a black-box approach for various reasons. The quality of code available for free download is debatable, and if you don't know what's happening under the covers, you could easily obtain incorrect results. Anyone using programming needs a solid foundation. Otherwise, their output may be wrong or the analysis poor. So, if you use the copy-paste approach, it's important to be mindful of the codes you borrow. By the end of this book, you will have an idea of each line of code, but this book is not meant to make a Python developer out of you. Instead, you are a psychologist who needs to learn how to analyse their datasets with machine learning using Python.

I believe that I have included almost everything you need as a psychologist to analyse your quantitative data and probably the only book required for many readers. But that said, data science is a huge subject. Therefore, I have made this book simple, short and easy to understand. In that effort, I removed many advanced-level topics that are otherwise important (e.g. how to analyse texts or video data). So, for some people, this book is not the last book but the first steppingstone.

Specifically, you will learn:

1. How to code in Python 3?
2. How to collect the data?
3. How to make it ready for processing?
4. How to implement a machine learning model?
5. How to explain the predictions of the machine learning model?
6. How to interpret all the results?

1.1.2 Who Am I?

Chandril Ghosh is a UK-based chartered psychologist and is currently working as the Lecturer in Clinical/Counselling Psychology at the Bath Spa University. I am also a committee member of NI British Psychological Society. In the past, I completed my BSc in Psychology (honours) from Kolkata and then MSc in Psychology (Clinical) from Bengaluru. After that, I flew to the United Kingdom to finish my PhD in Psychology at Queen's University Belfast. For my doctoral dissertation, I used innovative data analysis techniques such as machine learning, network analysis and Jaccard's coefficient to analyse patients' narratives about their lived experiences of mental illnesses. In addition, I attempted to contribute to the descriptive literature on psychopathology. In addition to research, I was also associated with teaching machine learning using Python to the students of psychology, business analytics, economics, human resources and management at Queen's University Belfast. I also run my online courses for students from over 55 countries. If you want to enrol on my online courses, then you can get enrolment information from my website (https://chandril.org.uk/) or just drop me an email at ghoshchandril@gmail.com.

1.1.3 How Did I Get There?

I get this question a lot. What the person really wants to know from me is, "Can I do it too?". Let me try to answer this here. As you might have noticed, my formal education is in psychology. I never took a university degree in computer science or data science. So much of my knowledge was developed through self-learning using the resources on the Internet. Of course, it did not happen overnight. In fact, it took me years to just get the basics right. Also, I was fortunate enough to have met Richard Joltes on a social media platform, and he was kind enough to act as my mentor numerous times. He's a great guy who has 35 years of experience and works at Harvard University as the Adjunct Faculty of Information Systems. From time to time, I have also received the support and advice of my peers like Prof Deepak and Meenu from the computer science department, which has helped me shorten the path of self-learning. Prof Deepak and Richard have also peer-reviewed this book. To summarise, my learning track consisted of self-study and occasional mentorship from some very talented people.

As you are reading this, I assume you have a similar background in psychology and have never had any prior coding or data science experience. That's perfectly fine. In fact, that's what I have written this book for. As said before, the self-study path took me about 4 years, and as you can imagine, it was not a linear route. There were times when I was stuck at one problem for days at a stretch. I wanted to write this book to ensure that you don't have to spend those painful times and reach the outcome sooner than I did. So, I condensed much of what I learnt in the past 4 years into a SYSTEM that is easy to follow and quick to learn for a person with a psychological background. In gist, this book is meant to reduce my 4-year journey into 4 days of self-study for you, and by the end of this book, you will get the results without repeating my mistakes.

Don't get me wrong. There is no denying that making mistakes is part of the learning experience. The greater number of times you solve problems richer your experience will be. Completing this book will give you a booster start and will probably make your learning track more linear than it was for me. But I am confident that you will still make new kinds of mistakes, maybe different ones than the ones I did, and Python will throw errors. Those are important experiences and will help you develop your problem-solving skills. Remember, when there is a problem, there will be a solution. It might not be obvious at that moment, but if you take your time to try different combinations or browse the Internet for solutions, chances are you will find them and learn from that experience. To summarise, I can assure you that it is possible to learn it all by yourself (without attending a formal degree programme) and reach the knowledge I have gathered over these years (maybe much sooner and much beyond). This book is my attempt to compile and compress all the need-to-know topics on machine learning a psychologist needs to know to analyse their quantitative data for academic research.

Bonus

Do I need to code, do maths or learn the theoretical backgrounds?

Data science is huge. Depending on your needs, you can learn data science at different levels of depth. Some people shy away from learning with the fear of coding and maths. But let me clarify this misconception for you right away. There are many routes to learn data science today. There are routes on which you don't need to learn to code or do single math. So, do not panic. If you don't want to code, then you can use software like KNIME and Tableau. You do not need to code to get started with machine learning or artificial neural network. But remember, if you have big data, such platforms might crash, or your computer might get hung up. So, there are some advantages of learning to code. In this book, we will learn Python coding in the easiest way possible. But if you are not so much in love with maths, then no worries, we will not learn any of that maths associated with data science.

1.1.4 Who Is This Book For?

So, I wrote this book specifically for academic psychology researchers (quantitative). There are no prerequisites, but this book will help someone with an undergraduate degree covering the basics of psychology, statistics, research designs and methodologies. I am also assuming you have had basics of statistics. Even if you are not good at it or despise it, knowing the fundamentals will help. The amount of statistics learnt in the undergraduate curriculum of psychology is enough.

Machine learning can handle any kind of data. It can analyse both numbers and texts. So, both quantitative and qualitative researchers might benefit from this. To make this book short and simple, it only covers how to analyse numbers. For learning to analyse textual data, we need to learn Natural Language Processing (NLP). Much of my doctoral thesis utilises NLP. From my personal experience as a learner, I feel NLP should be the second step for absolute beginners. The first step should be to learn to analyse numbers which are way easier to get started. So, I have not covered how to analyse textual data in this book. Sorry qualitative researchers! But after completing this book, I promise it will be much easier for you to grasp NLP. Hopefully, I will write a book for you on text analytics soon.

1.1.5 Who Is This Book NOT For?

This book is not for people who are seeking to read a comprehensive book. It is for those who are looking for a quick start guide. You will find big fat books on the subject matter in the market. The idea of this book was to be as thin as possible to encourage more and more psychology researchers and students to get started and learn the subject as quickly as possible. In my experience, often fat books scare people away because it demands a lot of time, effort and patience from the reader. It makes no sense to dump a psychology researcher or student with a comprehensive book that is already slightly shaky about numbers (stat-/math-phobic). How many books have you started but never finished? I have a dozen of them.

On the contrary, thin books like this one attract me personally as a reader. It's easier to convince oneself to commit to it and successfully finish it. Once you finish it, you get some confidence to start. A brilliant big fat book often fails to achieve this vital task. In my opinion, the first step should always be small but challenging enough to make you feel confident by the end of the adventure. That is exactly how I wrote this book. To do so, I have removed all machine learning elements and Python, which I feel you will not need as a psychological scientist. For example, I have excluded the mathematical derivation or explanations from this book. I am also guilty of avoiding in-depth explanations of certain important concepts' workings because you will not need them in practice. Including those will only make you feel lost in technical details. In short, I have written this book to cover whatever you will need in practice to publish research papers. Finally, this book fits the interests of quantitative researchers more than qualitative researchers because we will learn to implement machine learning models on numeric data. In the end, if you already know the basics of data science and machine learning or seek a book on industrial applications of machine learning, then this book might not be the best one for you either. I wrote this book for absolute beginners based in psychology academia.

1.1.6 *What's Special You Get in This Book?*

My idea is to compile all my experiences in this book and present you with a shortcut to learning the basics fast and well. You do not have to repeat the same mistake or feel lost as I did. In a short and sweet format, we will learn how to analyse data using machine learning in this book in a language we are familiar with. It's brain-friendly and easy to understand. Readers who want a result will learn a new skill and experience a change in their lives, and this book intends to guide them to get there. No filler, no trial and error and no time wasted.

There is a reason why I took the time to write this book. Many books, blogs and free videos are already on machine learning. They are great. I was self-taught using those resources. This book is not a competition, and I highly recommend exploring those resources after completing this book. But there are rarely any curated resources on machine learning for our tribe and us. I mentioned the excellent resources above, but most of them come from computer scientists and engineers. They are no doubt some of the most brilliant minds on Earth. But there is a serious problem. They are not psychologists themselves. They speak a different language, use jargon that we never heard and worse, use the same word in a different context. I learnt from these people, and my first-hand experience was that I was lost and confused a dozen times. Also, there is another big difference. Those computer scientists and engineers were focused on industrial applications of machine learning. They had little or no idea about the nature of problems we deal with in behavioural and social sciences. While they taught me how to use machine learning to build a SPAM email filter, this lesson was little useful for academic contribution for a psychologist interested in people's mental health experiences. Again, it took me a long time to realise this problem and, after that, find a solution to use for my research work.

1.1.7 *So, What Does This Book Have?*

If you go back and look at the contents of this book, it should give you a fair understanding of what's in this book. For the sake of simplicity, we will learn this in three steps, each building upon the previous step. The three steps are as follows:

- Step 1: Learn to code in Python
- Step 2: Learn to refine the raw data
- Step 3: Learn to analyse the data with machine learning

1.1.8 *How to Best Make Use of This Book?*

As you progress with each of the following topics, do these two activities:

1. Do what I just did during the code demonstration on your computer: mere replication. Feel free to use my codes on your own dataset. Even when you are using my codes (which worked well for my data), you might get errors while running the same for your dataset. I want that to happen. Read the error message carefully. Jupyter will give you a hint on what part of the script it is unable to understand. For example, there might be issues with the format of your dataset, such as if you try to read an Excel file using the "pandas.read_csv", which is meant for reading CSV files. There are innumerable possibilities of getting an error. I cannot list them all. But I would be sad if you don't get errors because getting errors is part of our daily life. Unless you learn to resolve them from an early stage, you will probably not learn Python that well. Suppose you were trying to learn to swim, then only when you drown a couple of times in water you would learn how to swim.
2. While reading each topic, ask yourself, "How will you apply what you are learning?" Keep asking this last question religiously after every topic. Write your thoughts on how well you apply what you learnt in your project. This will strengthen the neural connections in your brain, and learning will be better and more memorable.

Bonus
In Python, getting an error message is not a sin. It simply means that Python cannot comprehend or understand what you are saying. You might feel the same as the experience when you are just learning a new foreign language and then trying to speak in that language with a native speaker. There will be such bonus points time and again throughout this book to give you crucial information.

1.2 Types of Research Studies

This book assumes that the reader already knows research types, designs and methodologies. So, we will not cover them in depth. Instead, the purpose of this part of the book is to give you a refresher on some of the main types of research based on their aims and to show a couple of ways in which you can implement machine learning in them.

For starters, there are two major cultures within the mainstream research: data modelling (explanation-focused) and algorithmic modelling (prediction-focused). Followers of the data-modelling culture conduct **explanatory research** that is concerned with estimating the true parameters of the process, while the followers of algorithmic modelling conduct **predictive research** that is more interested in finding an algorithm that results in the same outputs as this process given the same inputs. There is also a third type of research in psychology for the topics that have not been studied much in the past, and the literature is scanty. For such lines of investigations, we can perform **exploratory research**. In this part of the book, we will take a quick peek into each of these three in the context of machine learning. Figure 1.2 presents the three common variations of research aims.

Spoiler: For explanatory and predictive research, supervised machine learning is used. For exploratory research, **un**supervised machine learning can be used. We will cover this later in-detailed.

I recommend you read this portion of the book now and once again after you are done with the book. The rationale is that by reading the chapter on machine learning in this book, you will forget most of what we learnt here. So, after you go through the lessons on machine learning techniques, parts of this chapter will make more sense, and it will be easier for you to implement the learnt lessons.

1.2.1 Explanatory Research

Broadly, psychologists are interested in studying people's thoughts, emotions and behaviours. Some are interested in developing interventions (e.g. smoking cessation) and typically use experimental design to study such topics. Others are more interested in finding relationships between different variables, including predicting a target behaviour (e.g. suicide attempt) from a variety of factors (e.g. symptoms). They might use a survey design for that. Let's first talk about the studies that use experimental design.

Experimental Design

Let's say we are interested in studying whether a particular intervention (e.g. Cognitive Behavioural Therapy) changes a target behaviour (e.g. smoking). We prefer a pure experimental design. First, participants are randomly chosen from the population. Then the subjects are evaluated on the variable of interest (before the intervention). For example, our variable of interest is *the number of cigarettes smoked daily*. We would simply ask the subjects to report how many they smoked in the past 24 hours to collect the data. Afterwards, we would expose them to the intervention (e.g. a smoking cessation programme). After some time, we would ask them how many cigarettes they smoked now. We would then compare the mean of cigarettes smoked before and after the intervention. We would probably use the t-test to compare those means. If our program was successful, then a significant drop in cigarette smoking after the intervention. This is called a one-group pre-test-post-test design because there is only one group of the sample.

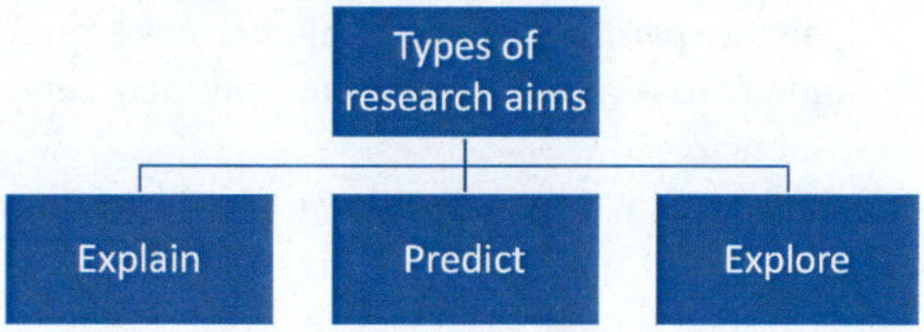

Fig. 1.2 Different aims of research studies

There are other methods, as well. For example, two (or more than one)-group design is another method where one group is administered the intervention (called EXPERIMENTAL group) and the other group is not administered the intervention (called the CONTROL group). Three variations of such designs are:

- **The pre-test-post-test control group design** is very similar to the example above (i.e. one-group pre-test-post-test design). The only difference is that we have two groups of subjects in this case. We assigned some to the control group and the rest to the intervention group. Then we estimate their smoking before and after a certain period. During this time gap, the experimental group receives an intervention (e.g. smoking cessation program). The control group either receives no intervention or placebo intervention (e.g. sugar pills). The difference in smoking is then compared between the two groups. If the intervention was successful, the experimental group should see a significant drop in cigarette smoking. The control group should not have a significant difference between its pre- and post-tests. You might use a two-group Analysis of Variance (ANOVA) between the treatment and control groups.
- **The post-test-only control group design** is arguably more commonplace in behavioural science because you often do not get to measure or test people before an intervention. Especially if you are conducting field experiments such as the Petrified Forest Study (Cialdini et al., 2006), so, in such cases, we do not have the pre-test. Instead, the effect of the intervention is measured as the difference in the post-test scores between the two groups.

Let's explore this further using the Tipping study's example (Strohmetz et al., 2002). In summary, the experiment studied reciprocity. The idea was simple. Some of the diners in a restaurant were offered a small piece of chocolate along with the check. Others were provided only the check. The study found that the diners who received the chocolate tipped more than those who did not receive a candy. What I want you to do is make a note of the research design.

Note

1. The independent variable (intervention) was giving chocolate along with the check. The control group diners were not given chocolates. The experimental group diners were given.
2. The dependent variable (target, outcome) was the amount of money the diners gave as the tip.
3. The setting was a restaurant (field experiment) and diners probably never knew they were part of a research experiment.
4. No pre-test was done. This is obvious because the waiters or servers (who were also experimental accomplices) never met the diners before. Even if they had met, this was the first time they had been asked to do the diners' experimental task.
5. Subjects were randomly assigned to the control group and the experimental group.

- **Covariance design:** This design is similar to the pre-test and post-test designs mentioned above. The only difference is that the pre-test does not measure the dependent variable (e.g. number of cigarettes before intervention). Instead, we measure the covariates of dependent variables (e.g. age, occupation and so forth). We desire to study the independent variable's effect (the intervention) on the dependent variable (target behaviour). The covariates mentioned above might influence the dependent variable, making it difficult to claim if the dependent variable's change was because of the intervention or these covariates. Therefore, we try to control these covariates using such a design. We can use a two-group covariance analysis (ANCOVA) to analyse the data.

An important part of all kinds of experiments is randomly assigning people into appropriate groups (control and experiment). Random assignment is desirable because it lets us assume that groups are equivalent (before the intervention). This also means any differences found in the post-test between both the groups are because of the intervention. The likelihood of age, race, sex, and other variations influencing the dependent variable is minimised.

Now we can do this random assignment using a simple random assignment. But this method of assignment does not recognise subgroups within a population. We understand that group membership, such as age or sex, might influence the dependent variable's post-test results. To avoid such instances from happening, we can adopt three sampling methods:

1. **Stratified sampling:** First, divide the sample population into groups called strata and then into the experimental/control groups. If we divide the sample into sex (strata), then the researcher would pick an equal number or proportion of people (representative of the population) for the study. This ensures that every part of the population is represented.
2. **Cluster sampling:** Alternatively, we can divide the sample into groups ("clusters"). For example, we divided the sample into three groups. We can then randomly select one of the groups for control and another for the experimental group, using a simple lottery method. We will omit the remaining third group from the study.

Now, Where Does Machine Learning Fits Here?

We know that stratification and clustering are desirable. But how to select strata or cluster people in practice?

- Let's start with strata. Many traditional strata building approaches pick all or numerous variables (e.g. age, sex and income), which results in too many strata. Here's how a **stratification method** works, step by step:

  1. The researcher decides how they want to stratify the population. Let's imagine that there are ten people in your sample. You, the researcher, decide to stratify them in terms of sex: male and female. Note, here, the researcher decides on what basis (or covariate) to stratify, leaving room for bias. So, what if the researcher assume that "sex" would be appropriate for stratifica-

tion but in reality, "age" was more important for addressing this research question? Alternatively, some researchers choose all/many demographic variables for stratification. Still, that might generate too many strata divisions (which is not conducive to studies with a smaller sample). Fortunately, **machine learning** can make this process more systematic and easier (for review, Aufenanger, 2017). Ideally, we should stratify (Bruhn & McKenzie, 2009). We can use **random forest** and its feature importance functionality to discover which covariates contribute exactly to the dependent variable. For instance, we might discover that only income contributes to the dependent variable of age, sex and income. Then we can use it to stratify the sample based on the salary variable. There is no point in stratifying based on sex and age. The subsequent steps remain the same as we did in traditional methods.

2. In the next step, the researcher decides how many people in the population (n = 10) are male or female. Let's say five are male, and the other five are female.
3. The third step is to decide on your sample size. Let's assume your sample size is 6 (from the population size of 10). You can use Cochran's sample size formula to decide on that.
4. Fourth, it would help if you calculated the proportion of people from each sex. In other words, how many females and males should we pick from the population for the study sample? The answer is three females (= 6/10*5). The formula is simple: a sample size of the strata = size of the entire sample (i.e. 6) / total population size (i.e. 10) * number of people in strata from the population. So, you have three males and three females in the sample. That's all for stratification.
5. In the last step, you must do simple random sampling in each stratum to assign the participants into control and experimental groups.

- Now let's look at clusters. Earlier with stratified sampling, we drew a random sample from each stratum (e.g. three from male strata and three from female strata). This is not the same for cluster sampling. In this case, a researcher chooses and studies only the chosen clusters. Again, as with stratification, traditional approaches rely on the researcher to select the clusters. The reliance on the researcher risks bias and potential confounding of the causal relationship. Instead, we can use **k-means clustering** to cluster the sample. There can be more than one covariate. Being data-driven, instead of the researcher (human) driven, the groupings are lesser prone to human errors. The k-means clustering result will tell us which people in the sample share similar qualities. Accordingly, we can tag these people in the same group with labels such as A, B, X, Y and so forth. This allows us to carry the research on ONLY the people belonging to cluster A.

Later in this book, you will learn these machine learning techniques, such as feature importance (Page: 130) using random forest and k-means (Page: 143). Integrating the random forest algorithm and k-means clustering algorithm (from machine learning) reduces human biases, increases the study design's validity and adds confidence to the data-driven findings. However, not everything is a piece of good news. It might add some extra work as you have to integrate these algorithms in experimental designs to stratify and cluster.

Research designs can be creative and unique. I do not know what yours is. But chances are, for experimental design, you might have to collect observational data and conduct a pilot experiment (an additional step) for stratification and clustering. If you have the time and resources and if it is within your research scope (certain forms of field experiments might make it a bit more difficult), integrating machine learning might be the way forward. Otherwise, stick to the good old ways of relying on the researcher's intuition and try to attribute to literature to conduct such clustering or stratification as much as possible. You can probably integrate machine learning algorithms with little or no extra data collection in survey designs. We will explore the scopes of machine learning in the survey designs below.

Non-experimental Design

This is probably one of the most favourite research designs for social scientists and data scientists alike. It includes survey designs and, more recently, social media research designs.

Survey designs: It's probably the design you are most familiar with. Maybe, you conduct survey research for your dissertation or have been a subject who was surveyed. Either way, if you had your behavioural or social sciences schooling, chances are you are familiar with it. But just in case you are not, in survey designs, we survey people (using scales or questionnaires). Then we enter the data on the spreadsheet. Usually, rows represent the participants, and the columns represent each variable. Finally, we perform statistical analyses such as correlation and regression to answer the research questions. We can extensively use almost all forms of machine learning algorithms to do more with such data. Machine learning opens up entirely new possibilities with what inferences you can make from the data analyses. For example, we can predict dependent variables from independent variables. In cases where you have only independent variables (and your goal is exploratory), machine learning can help your group or cluster people from that dataset.

1.2.2 Predictive Research

The above discussion was based on explanatory research. In this part, we will learn about predictive research.

Currently, studies in psychology are primarily concerned with explaining the causal mechanisms that give rise to behaviour or experience (data modelling, explanation-focused culture). Such explanatory research has helped us develop intricate theories of psychological mechanisms. For example, it has helped us answer questions like why individuals who have been abused as children tend to be at higher risk for developing psychopathology as adults. Of course, we use conventional statistics, and the results are evaluated either using "goodness of fit" between the statistical model and the sample data (to determine if the sample data represents the data that would be expected to find in the actual population) or using whether the sizes and directions of certain regression coefficients match with what is implied by different theoretical perspectives. But such assumptions and explanatory studies generally have some serious concerns.

The problem with explanatory studies is that they explain the causes of behaviour, leading to the development of mechanistic psychological models that hold theoretical appeal but rarely display a meaningful capacity to predict future behaviour. One might argue that we do "predict" in explanatory studies. For example, there are studies which are titled *variable X predicts variable Y in people of a specific demography.* Such claims have been based on the assumption that the model that best approximates the psychological processes that produce an observed behaviour or experience is also the one that best predicts future behaviour/experience. While explanation and prediction are philosophically intertwined concepts, from a statistical standpoint, it is simply not true that the model that most closely approximates the data-generating process will generally be the most successful at predicting real-world outcomes. Due to a phenomenon known as overfitting, a biased, psychologically implausible model can often systematically outperform a mechanistically more accurate and more complex model. So, while it is true that in explanatory studies using regression equations, we can readily generate new predictions simply by plugging in the corresponding values of X1 and X2 for newly observed cases. As a result of such plug-ins, we obtain a coefficient of determination of, say, $R2 = 0.7$; then, we can report that it can "predict" 70% of the variance in the target variable using a set of predictor variables. So, one might argue that we are claiming that to make reasonably accurate predictions about the target variable of a random person drawn from the same underlying population. But the problem lies in the inference that the coefficients (parameter estimates) obtained in the sample at hand will perform comparably well when applied to other samples drawn from the same population. Such a fitted model will almost invariably produce overly optimistic results because the relationship between variables in any sample is always influenced in part by sampling or measurement error – which is not shared with other data samples drawn from the same population. Such studies focusing on good model fit or theory-congruent explanation can reduce the likelihood of generating good predictions. Furthermore, many published papers in psychology do not replicate well when the same experiments and analyses are independently conducted later ("replication crisis").

Fortunately, explanatory studies are not the only type of studies, although most commonly, that is what statisticians and psychologists who use statistics pursue. The alternative is predictive research. Practitioners of this line of empirical research are more concerned with forecasting future events or behaviour. So, for example, a researcher might collect school grades and standardised test scores to predict the future annual income of the person. So, while explanatory research focuses on "explaining" the behaviour, predictive research focuses on "predicting" the behaviour. As simple as that.

1.2.3 *Exploratory Research*

Exploratory research design is used for investigating a problem that is not clearly defined. Consider it to be a form of preliminary research that requires to be done to define problems and suggest hypotheses. One way to conduct exploratory research is to group a set of people (or entities) so that members of each group are more similar to each other than they are to the members of other groups. For example, if you have a sample of people that you wish to subgroup based on their illnesses, then probably you can form three groups: one group will have all the patients diagnosed with diabetes, the other group has all the patients with cancer, and the third group might have all the people with neither of these two conditions (diabetes and cancer). This is called clustering. In this book, you will learn to implement machine learning models to do clustering and explore the hidden patterns within the dataset as well. We will use unsupervised machine learning models (Page: 136) such as **hierarchical clustering** and **k-means clustering** to do such grouping of people based on their similarity. After the clusters are formed, you can study the mean scores of each variable among the people of each cluster to find in which manner the clusters are different from one another.

1.3 Data

What Is Data?
Data is information. Information can be anything. Your name, age, sex, the city you live in, everything qualifies for being called data. Imagine if you are visiting a new city. The more information we have about it in advance, the better, safer and smoother it is to travel around it. This is the whole idea behind being data-driven. As mentioned before, data can be anything. Your name can be an example of textual data. Your age is numerical data. The photo you shot of your favourite restaurant is image data. The video of your wedding ceremony is, of course, video data.

1.3.1 *To Collect or Not Collect Your Own Data*

Psychologists are arguably more inclined to collect primary data. At the same time, other professionals, such as data scientists (and economists), are more inclined to use secondary data. There is a valid reason for each. When you are a psychologist planning to use data scientist tools, it is probably important to understand why data scientists use secondary data. The reason is, usually, machine learning algorithms use a huge amount of data. It's economical to get the data collected by someone else from a data scientist's perspective. As a researcher of behavioural and social science, too, in my opinion, your first hunch should be to use the data that already exists. It just saves you from a lot of pain and effort. You can use the freed-up time to focus on the other things that matter to you and your research. The good news is that much of these available large-scale secondary data are free of cost. But you might need some training and some more time to get ethical approval on your intended use of the data. An example of such training would be the **Safe Research Training** (https://www.ukdataservice.ac.uk/get-data/how-to-access/accesssecurelab/train.aspx), especially if you are based in the UK and seek to use the UK Data Service.

I recommend you look at the **UK Data Service** (https://ukdataservice.ac.uk/). Examples are:

- The 1970 British Cohort Study (comprises 17,000 people of Great Britain in a single week of 1970)
- The Millennium Cohort Study
- The United Kingdom Household Longitudinal Study
- The National Child Development Study
- The European Quality of Life Survey (cross-sectional dataset)

When to Collect Your Data?

There are times when the dataset does not have the variables which you need. Those are the times you should opt for collecting your data. Let me give you a personal example. I was working on mental health research for my PhD. Currently, much of our mental health studies and clinical practice are guided by the diagnosed guides such as the *Diagnostic and Statistical Manual of Mental Disorders* (DSM) and the *International Classification of Diseases* (ICD). My thesis supported the growing literature criticising the DSM and ICD for low reliability and validity. But as I said before, they are frequently used worldwide even while conducting the surveys. Understandably, when I checked some of the datasets mentioned above, I found that most of them used mental illness diagnoses and had no mention of individual symptoms. I was hoping to get a dataset that had questionnaire-item-level variables (e.g. individual symptoms), and these datasets proved to be useless to me. So, I had to gather my own data. But the point is that you will find secondary data in line with what you want every time. Suppose you don't want to compromise your research questions and directions based on available data. In that case, you might have to collect your data. More painful but worthwhile. We will learn more about the sources of data on page 17.

I must reiterate the fact that much of what we will learn in this book relates to numerical data. But suppose you are a qualitative researcher working with interview data or other forms of textual data. In that case, I must tell you that you can use machine learning algorithms to analyse texts. For example, there are websites like LiveJournal (https://www.livejournal.com/), where people worldwide come to post their views and experiences. Then there are e-commerce sites with reviews about products from consumers. We can scrap the websites and generate a database of these narrations. Imagine an Excel sheet with each row representing one individual and what they said about a product, themselves and the like. As you might imagine, these are textual data. Qualitative researchers might find this avenue interesting and is mostly used to discover what people say about something or someone on the Internet.

I can give you a personal example of when to collect your own data. During my PhD in psychology, I had scraped peoples' mental health experiences from the LiveJournal forum. The idea is to understand psychopathology's experiences from patients' first-hand reports. So, I mined the collected data and ran it through algorithms to figure out how the reported symptoms relate to one another. I used a network graph on text-based data. Analysing textual data is also referred to as *Natural Language Processing*. I consider it the second step for students at your stage and beyond this book's scope. Let me assure you that this book is the first step to cover all the prerequisites if you are excited about it. I hope to write the next book on Natural Language Processing with Python for psychologists, finish this book sooner and keep in touch with me for the announcement (https://chandril.org.uk/).

1.3.2 Where to Get the Data From?

We can collect data ourselves (primary data), we can use the data collected by someone else (secondary data), and finally, we can use the data compiled by someone (tertiary data) [Fig. 1.3]. Take the example of visiting a new city. Well, you can go for yourself, navigate the city and explore different eat-outs and restaurants yourself. But you can also ask a friend or colleague who has been in that city before recommending you the best restaurant or reading a food blog.

This book will focus on primary data because psychologists in the field gather more often. That being said, I will also discuss the secondary sources of data and provide you with the necessary links.

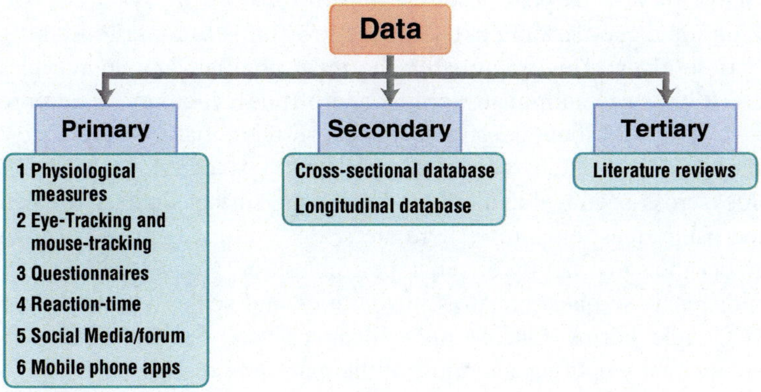

Fig. 1.3 Different sources of data

Sources of Primary Quantitative Data

- **Physiological measure:** If you have a good budget and/or your lab has resources, you can use physiological measures to investigate the biological correlations of behaviour. There are several options. We can measure skin conductance responses (SCRs) using electrodermal activity (EDA), cardiovascular activity using an electrocardiogram (ECG), muscle activity using electromyography (EMG) measures, hormone levels using blood samples or saliva samples, electrical activity in the brain using electroencephalogram (EEG) and event-related potentials (ERPs) and study the functional activity in the living brain using functional magnetic resonance imaging (fMRI) and positron emission tomography (PET). For a quick review, refer to the chapter (Vanman & Philipp, 2019).
- **Eye-tracking software:** People look at the computer screen as an important source of behavioural data because it tells us where they are paying attention. It is a simple setup. First, we put a participant behind the computer screen and pointed a camera (can use a webcam) at her eyes. Then, we can show the person images on the screen, while the eye-tracking software records the eye movements. Commercial options are expensive and can ask for more than a couple of thousand pounds. But fortunately, there is an open-source version for eye tracking in Python (http://www.pygaze.org/).
- **Mouse-tracking software:** How people interact with specific content on the computer screen (e.g. a web page) can lend invaluable insight to user-design professionals. Behavioural scientists can benefit too. You can use a computer mouse-tracking method for assessing real-time cognitive processing in psychological tasks (Freeman & Ambady, 2010). You can use the free available software (http://www.mousetracker.org/). But note, it's for specific psychological tasks on the screen. What if you want to study more real-world scenarios, such as what users are doing on a web page or blog? Then you can opt for the commercial mouse-tracking software. Those are more targeted to help businesses decide how to optimise their site usability and design. But you, as a behavioural scientist, can use it too for your purpose. Many such paid software are available, but if you are looking for a free-tier option, try Mouseflow (https://mouseflow.com/).
- **Questionnaires:** This is probably the most popular data source in social sciences. It's easy to administer (online and offline), economical and probably the one way to know about constructs like self-esteem that are otherwise not visible or measurable through other data sources. Most of the large secondary data sources, cross-sectional and longitudinal, were survey studies. The data was collected using questionnaires. There are scales and inventories but imagine them all under the same family of self-report measures. Nowadays, many researchers administer these questionnaires, inventories and scales via the Internet. I have used Google Forms. One of my colleagues used SurveyMonkey. It depends. There are many options, and most of them are free of cost.
- **Observation:** While in the social sciences, questionnaires are popular, in behavioural sciences, observation reigns. It can take many forms. For example, in the Petrified Forest study, the researchers marked and placed 20 pieces of petrified wood on the path (making them accessible to the potential thieves). After a while, they counted how many of those 20 were stolen from that path. The result was

compared for all three paths. Now, remember the Tipping study. The researchers observed how much money was tipped by those diners who received candy and compared the amount tipped by those who did not receive candy. If it slipped out of your memory, or in case you skipped these two studies, refer back to page 12.

- **Reaction times:** It is popular among experimental psychologists and calculates how quickly an individual can respond to a particular stimulus. This has real-world implications. One example is if danger sign X takes a higher reaction time for the drivers than danger sign Y. To ensure the safety of road drivers, sign Y will be chosen because sign X takes drivers more time to realise a danger and push the brakes. There are several free online measures of reaction time. You can also use the software PsychoPy (https://www.psychopy.org/) for reaction time. It is open source and widely used by experimental psychologists and neuroscience (Peirce et al., 2019). In addition to reaction time, this software can conduct many experiments for you. So, if your research requires a computer-based experimental task, this deserves a shot.
- **Internet:**
  - **User-generated content:** Obtaining qualitative data through interviews is a painstaking job. An alternative is gathering the post contents on what people share online. People post tons of such content daily about their views, experiences and the like. Imagine if you can download all of them with a few mouse clicks and then analyse them for similarities and patterns and draw insights about the phenomena in question. For me, I wanted to understand what people with mental illness experience. So, I went to an online forum (www.livejournal.com), searched for such mental health communities and scraped the content to download it on my computer in an excel format where each row represented a unique individual they had said. The total sample size was somewhat around 10k people. It took barely a few seconds to gather that data.
  - Finally, I used data-analytic algorithms to analyse and find the mental illness's hidden patterns from patient-reported experiences. So, I did analyse qualitative data (collected from the Internet) using quantitative methods. You can read how I did it from the published research paper (Ghosh et al., 2021) to learn and adopt a similar method for your own research. Read the following study for a systematic review of user-generated content (Naab & Sehl, 2016). **Note:** It's important to consider data veracity (i.e. trustworthiness) and bias. For example, some topics are spoken about on social media more often than others. For example, people will talk more elaborately and frequently about product reviews or their success stories than sexual dysfunction. You need to imagine if your research area is sensitive and if people feel shy to write about it online. In my case, mental health is sure a sensitive issue. But LiveJournal allows users to post their experiences anonymously. So, that reduces instances of shying away from certain sensitive topics. But even when anonymous (within mental health discussion groups), I found people spoke more about their anxiety and depression than sexual dysfunction. If you intend to use my method, it will help to be mindful of the study topic. Likewise, suppose you're looking at social media or online community data. In that case, it's automatically biased because it's put there by people who are comfortable posting information online, have access to a computer and are savvy enough

to find such sites. **To sum up, social media and online forums can be invaluable qualitative data sources.** You can use machine learning and other data-analytic methods to draw inferences. But this topic, in my opinion, is for advanced learners. Hence, I will keep it out of this book. You can read my research studies for curious readers and dig into Natural Language Processing or text-based machine learning. If you need help or consultancy, then feel free to email me. This book will not cover text data, but it's good to know the possibilities at this stage.

- **Digital footprints:** People say they intend to do something but do something completely different. The focus of behavioural scientists is often peoples' behaviour in action. Therefore, what people do on their phones and computers can be a gold mine (source of data) for behavioural scientists.

  Tracking with mobile phone apps: You can use special mobile phone tracking apps to collect behavioural data from your research participants and learn about their mobility patterns (Geyer et al., 2018), how much time they spend and on what (Deng et al., 2018), social interactions and so forth. To review such studies, read the following study (Harari et al., 2016).
  Tracking computer: Do you know so many employers monitor their employees' computer activities using tracking software? It helps them see how paid employees use their time at work. There is no reason why behavioural scientists cannot use the same. I have tried ActivTrak (https://activtrak.com/) for my personal use. I suggested ActivTrak because it has a free option (which I use). I tracked my computer activity to see how I used my time over a month and with the insight to increase productivity.

Bonus
You might experience some difficulty in recruiting subjects for your survey (or experiment). Why not put little behavioural science strategies to use? All you need to do is stand in a crowded public spot such as the university canteen (or wherever your target subjects go to hang out). Before asking potential candidates to fill the survey for you, ask them this: *Do you consider yourself helpful?* Most people consider themselves helpful and are likely to say YES to your question. What you did was induce a self-image for the person. The person is more likely to behave consistently with that image of "being helpful" by complying with the request. If you ask the person to help you with the survey, they will likely agree. However, don't forget to mention the selection bias under the limitation section of your study. This methodology selects people willing to respond to your query, which also needs to be considered in the study design. However, this strategy works well. A real study was done on the same methodology (Bolkan & Andersen, 2009). Among participants exposed to the helpfulness image induction (experimental group), about 77% agreed to help the experimenter participate in a survey.

On the other hand, for people who were not asked this question in advance (i.e. "Do you consider yourself helpful?"), only 29% of people (part of the control group) became willing to help. I understand a lot of us to conduct these surveys online today. I cannot imagine this strategy not working in an online environment. You can try posting the survey link with a simple eye-catcher line: *Do you consider yourself helpful? Take my survey to help me out*. With little scepticism, I guess it might work there too.

Another tip about gathering data from behavioural science literature is its importance to check the wording of the questions or rating scales. Selective attention elicited by directional questions can lead to biased responses (Kunda et al., 1993). So, instead of asking "*to what extent you are satisfied with this brand?*" consider asking "*how satisfied or dissatisfied are you with this brand?*" to get unbiased responses.

1.3.3 Ways in Which Data Is Divided

One way to divide types of data is in terms of their format. For example, data can be numeric, text, sound, image and video. We will talk only about the numeric dataset in this book. Even when there are textual data values such as male and female, we will convert it into a numeric form such as 1 and 0, where one stands for male and 0 stands for female (for example). Another way to divide data is in terms of measurement scales. There are four types:

1. Nominal (categorical) data (e.g. male/female)
2. Ordinal data (ranking data such as socioeconomic class and satisfaction)
3. Interval data
4. Ratio data

While analysing quantitative datasets in machine learning algorithms (as you will find later in this book), it is important to determine whether the data (of target/dependent variable) is numerical or categorical. Figure 1.4 presents the different types of data.

1. The numerical data consists of continuous (e.g. age) and discrete (e.g. the number of close friends) data.
2. The categorical data consists of nominal (e.g. sex, hair colour) and ordinal (e.g. low mood rating) data.

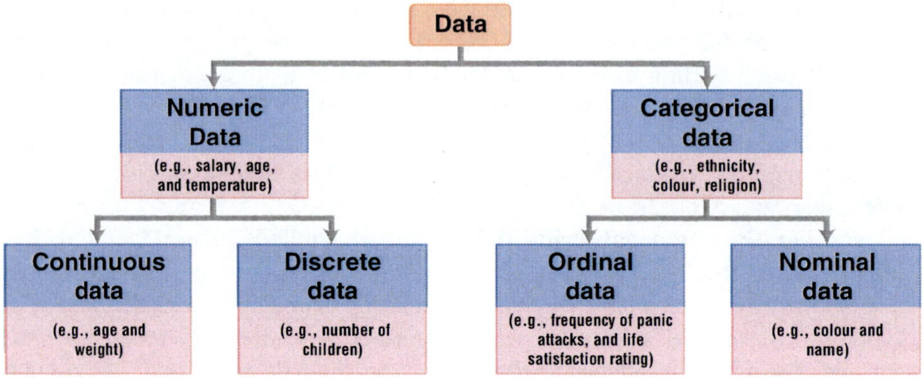

Fig. 1.4 Different types of data

1.3.4 Five Lessons

- In machine learning, you will often come across binary data. It falls under nominal (e.g. sex = male/female) and is usually represented by 0 and 1 or 1 and 2. Throughout this book and beyond, you will find machine learning models attempting to predict categorical variables (often in binary form). For example, predicting whether the tumour is malignant or not (an example of binary target data) can be an immense application. However, it's an application in the mental health domain that has been somewhat abused, in my personal opinion. It hurts to see researchers predicting heterogeneous syndromes such as depression and somewhat arbitrarily dividing participants as depressed or not depressed. Such studies are plenty and somewhat misleading for several reasons. If you are curious about this, please read my open-source research paper (Ghosh et al., 2021). You might not be a mental health researcher, but what I want you to learn here is to think twice before converting a continuous variable into its binary form. The machine learning model might provide you with highly accurate predictions on that. Still, suppose the model's variables are theoretically dubious or do not make sense. In that case, such a perfectly accurate machine learning model is useless. Likewise, the accuracy and usefulness of machine learning models are based on the quality of the data they're trained on. So, put trash data and variables into the machine learning model. It will produce trash results (even if it shows highly accurate).
- The machine learning algorithm model will change depending on the target variable's type of data. For example, if we have numerical data to predict, then we can use random forest regressor. But if the target data is categorical, then we need to use random forest classifier. As you can see, while the random forest algorithm can remain constant, the version of whether regressor or classifier differs. Right now, I do not expect you to understand what random forest regressors are and how is it different from a random forest classifier. We will cover them in the later chapter. For now, note that the type of data in the target variable is

important in choosing the type of algorithm. It's simple. It's either regressor (for continuous target such as score on the scale of low mood) or classifier (for categorical target such as depressed/not depressed).

- **What type of data do you think Likert scale ratings are?** In psychological research, we often use questionnaires and rating scales. What type of data do you think is the rating on an individual item of a questionnaire: categorical (i.e. ordinal) data or continuous data?

Let us suppose that our **target variable is an item** from the questionnaire. If it is a questionnaire for depression, then the item would represent one symptom (let's say, suicidal ideation). So, if we are attempting to predict that specific symptom as a target (e.g. frequency of suicidal ideation), then whether it is a categorical or continuous one is an important decision because it would dictate the choice of algorithm (i.e. classifier or regressor). If we choose the wrong algorithm, then our findings will be flawed. So, what is the nature of Likert-type data?

The answer is that it is an ordinal variable but can be used as continuous data. For specifically, it is called the *ordinal approximation of a continuous variable*. Confused? Give me a moment. Let me explain this. The Likert or ordinal variables *with five or more categories* can arguably be used as a continuous variable (Johnson & Creech, 1983; Norman, 2010; Sullivan & Artino, 2013; Zumbo & Zimmerman, 1993). To sum up, you can *use regressor for predicting item-level variables*.

- **Should we sum up the items to get a total score?** There are plenty of research studies opting to sum up all the items in the questionnaire and get a sum total. An example is that of a depression questionnaire. The common approach is to total all the items' responses (each representing a symptom). Suppose your target is the sum total of several items (or variables), such as a depression severity score. We can safely consider the summed score as continuous data in that case. But I must err on the side of caution with such an approach. Such a method of summing up the items to find a total score is valid if the construct (e.g. depression) is a single condition. All items (e.g. symptoms such as suicidal ideation and low mood) are equally good severity indicators. However, in many instances, this is not the case. More often, each item (representing a variable/symptom) is a different phenomenon and contributes to the construct (e.g. depression) to different extents. Therefore, the findings from the studies have questionable validity. Suppose you are curious to investigate the concerns by summarising the items to find depression severity. In that case, I recommend you read the following work (Fried & Nesse, 2015).
- **Other kinds of data:** We spoke about numerical data throughout this discussion and will speak more in this book. But just to let you know, machine learning practitioners are not limited to numbers. They can also work with any data format, including texts, video, images and audio. In the context of behavioural science, one popular example of using images as data would be the analyses of brain images using machine learning (e.g. Barranco-GutiÃrrez, 2020) and deep learning (for review, Lundervold & Lundervold, 2019). But for the sake of an easy grasp of the fundamentals and simplicity, we will stick to numerical data.

1.4 Statistics: A Refresher Before Getting into Machine Learning

Learning outcome: After reading this portion of the book, students will be able to *get an overview of the basics of statistics* to relate the similarities and *differences with machine learning and interpret machine learning results better.*

Both statistics and machine learning are methods of data analysis. There are many similarities. For example, you will find the term "regression" in both subjects. Likewise, one might argue that machine learning models represent a software-based approach to statistical analysis. For example, k-means and random forest existed prior to the invention of computing. We've just taken all those complex models and rendered them in software to make it easier to handle and process the data. Therefore, the line between them is fuzzy at times.

But there are major differences as well. One of the primary differences between statistics and machine learning is that "statistics draws population inferences from a sample, and machine learning finds generalizable predictive patterns" (Bzdok et al., 2018). In simpler words, inferences tell us "How psychological processes work?" and prediction tells us "What will happen next?" – both can be useful. However, the field of psychology has so far massively generated knowledge on inferences than prediction. As a result, while today we know a lot about the psychological processes behind suicide, we do not know when a particular person will attempt suicide. Learning machine learning in this book will enable you to do that (if you have the right data, of course).

Statistical models are based on certain assumptions about the data, such as multicollinearity and homoscedasticity. But in my opinion, machine learning doesn't rely heavily on these assumptions. It does to an extent, but machine learning practitioners are not as troubled by them as statisticians. We will come to realise that later when we learn about machine learning in the later chapter. For now, let us get an overview of statistics and some of its core assumptions and terminologies.

Statistics is important. Not just because it is fun or adds value to your claims but also because statistical skills are some of the most in-demand skills for a behavioural scientist. In other words, if you are good at statistics, it will help you get a job (even when you know little about other domains of the field).

Statistics rely on certain assumptions. One of the assumptions in many statistical models is that *the data of a particular variable is normally distributed*. This means if we collect a lot of data on a variable (I am speaking of the interval and ratio data) from a large sample, then the data distribution will be such that the majority of it will be concentrated on or near the mean. As a result, the number of data points will be lesser as we go away from the mean. A simple and familiar example would be to imagine the intelligence scores*. On the IQ test, most people will arguably score average, and fewer people will score too high or too less.

Imagine a *bell-shaped curve* where most peoples' IQ scores are centred around the mean and proportionally spread from the mean.

*Not related to machine learning or data science, but since I used the example of IQ tests, I consider it my responsibility to disclose that IQ scores are probably not really the measure of intelligence. Read more on this from the following study (Hampshire et al., 2012).

Coming back, so much of statistics assumes the property of normal distribution.

Now that I mentioned collecting the data from the sample (above), let me tell you what it means. Imagine you are conducting a study on car driving behaviour. It is worth mentioning that we cannot collect data from the entire *population* of car drivers. There are too many of them around the world. So, to play smart, the idea is to pick a group of car drivers to conduct the research and get the data from (Fig. 1.5). This group is called the *sample*. The idea is that if we study the sample, the result will be representative and generalisable to the entire population of drivers. Ideally, we should *pick the sample from the population randomly*. Still, it is not possible in many real-world settings, especially in behavioural and social sciences. We can adopt non-random sampling strategies such as convenience sampling in those cases. As the name suggests, the sample is picked from the population based on the researcher's convenience. For example, the researcher might want to interview whoever lives closer to the researcher. When sampling (*the process of selecting the sample from the population*) is not done randomly, we leave room for bias. As a result, our findings suffer from generalisability issues.

Next is **descriptive statistics**. After we collect the data, we describe it using descriptive statistics.

- **Measures of Central Tendency:** These help us find the centre of the data. Popular examples include mean, median and mode.
- **Measures of Distribution:** After we learn about the centres of the data, we need to find how the values are distributed around that central value. Measures of distribution, such as range and standard deviation, can help achieve the goal.

Next is **inferential statistics**. It involves drawing inferences from the data.

Often psychologists form hypotheses. These hypotheses are about what determinants or other factors influence the target behaviour. After forming the hypothesis, we need to put them to the test and see if they are correct. This is called hypothesis testing. Hypothesis testing is a form of inferential statistics.

Imagine an experiment where you hypothesise that mindfulness meditation reduces the frequency of anger outbursts. First, you need to test if it is right. So, you recruited study participants and randomly divided them into two groups: the experimental group and the control group.

Dependent variable: The frequency of anger outbursts in the past 7 days
Independent variable: The intervention, that is, mindfulness meditation (as "anger reduction intervention")

Bonus
As presented in Fig. 1.6, there are two other types of variables: **mediator variables** (which explain the relationship between the two other variables) and **moderator variables** (which influence the strength of a relationship between two other variables).

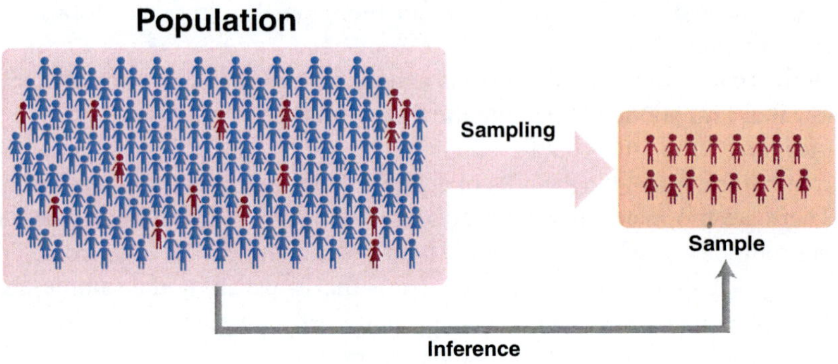

Fig. 1.5 The process of picking a sample from the target population

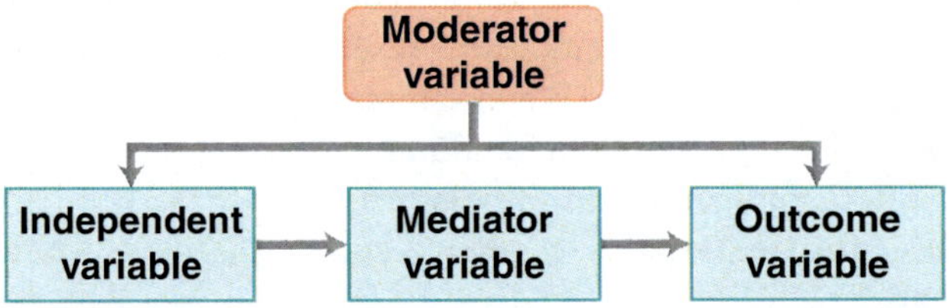

Fig. 1.6 The interaction of moderator and mediator variables on the relationship between the independent variable and the outcome variable

The experimental group received the intervention for anger reduction. But the participants in the control group did not receive any intervention. So now the question is if there is a significant difference between the means (i.e. frequency of anger outbursts in the past 7 days) of the two groups (during the post-test). If mindfulness meditation truly works, then the group of participants who received it should report a *significantly* lower frequency of anger outbursts. The statistical analysis to test our hypothesis is called **hypothesis testing.** Statistical tests such as t-tests enable us to compare the means of two samples and hence test our hypothesis.

Speaking of hypothesis, in this context, there are two types of hypotheses: null and alternative. In the example study mentioned above, the null hypothesis is there is no difference between groups (after the intervention). The alternative hypothesis will be just the opposite. So, we either retain the null hypothesis or reject it. After formulating the hypothesis, we then collect data to prove the null hypothesis false (so that we can say the intervention had an effect).

Note the wording here; we do not say that we accepted the alternative hypothesis. Instead, we say we rejected the null hypothesis. The focus on the null hypothesis is interesting.

This is similar to what happens in a courtroom. Yes, even the judge in the courtroom is doing a form of hypothesis testing (although not a statistical one). The law (in a lot of countries, including India) assumes that the person is innocent (until proven guilty). Accordingly, all the pieces of evidence presented in the court are directed towards proving the person guilty. Therefore, the courtroom's decision is either guilty or not guilty (instead of saying "innocent", which the person already is). Likewise, in statistical hypothesis testing, we assume that the hypothesis is a null hypothesis (until it fails to meet our pre-set statistical criteria). The decision of hypothesis testing is either to retain the null hypothesis (if it fails to do so) or reject the null hypothesis (the evidence suggests that this hypothesis is unlikely).

Take a more "academic" example. In a clinical trial that is being conducted to test whether or not a new psychotherapy reduces anxiety by 30%, the following will be null and alternative hypotheses:

- *Null hypothesis* (H_0): The psychotherapy reduces anxiety by 30%. $p = 0.25$.
- *Alternative hypothesis* (H_a): The psychotherapy does not reduce anxiety by 30%. $p \neq 0.25$.

Speaking of the statistical criteria for judgement, typically, we use the likelihood or level of significance and set it to 5% in behavioural and social sciences. Simply put, a p-value of less than 0.05 (typically ≤ 0.05) is statistically significant, while a p-value higher than 0.05 (> 0.05) is not.

- $p \leq 0.05$ (reject the null hypothesis, the intervention made a significant difference).
- $p > 0.05$ (accept the null hypothesis, the intervention made no difference).

But relying solely on p-values to make inferences has recently been challenged (Halsey, 2019). Instead, it is safer to report the **confidence intervals** and **effect sizes** along with the **p-value**.

Statistics are important, and learning them well will facilitate your data-analytics capabilities. But this book is not about statistics, so we won't dig deeper than familiarising you with statistics' essential terms and concepts. While I encourage you to pick a book on statistics first if your basics are not clear at this stage, I am not particularly worried because what we will learn in this book is somewhat independent of what you will learn in a basic statistics book. Take the example of hypothesis testing. There is no such hypothesis testing about the properties of a population (as mentioned above) in machine learning. Instead, we have binary classification. So, it matters little if you don't remember inferential statistics at this stage.

Bonus

While there is no hypothesis testing (as in inferential statistics) under machine learning, there are other kinds of "hypothesis" testing in machine learning. For example, when we run experiments using different combinations of variables in a machine learning setting, we are testing hypotheses. Likewise, when we change settings for a classifier (type of machine learning model) to alter the train/test dataset, we are again testing hypotheses. Naturally, many of these terms and concepts will not make sense at the moment. They will after you complete the chapter on machine learning. So, once you have read this book till the end, I recommend you to come back to some of these parts and re-read them. They will get much clearer.

Bonus

It can get a bit confusing whether you should frame your research problem as hypothesis testing (in which case you should go for statistical tests) or binary classification (in which case you need to go for machine learning) – I encourage you to read the paper published by Li and Tong (2020) for clarity on this matter.

From here, we will quickly overview the kinds of tests we have in statistics. First of all is the measure of association, called correlation. The measure is between −1 and +1, reflecting the degree of linear association between two (or more) variables. But just because the estimates of two variables vary together does not mean one is causing the other. Hence the popular saying "correlation is not causation".

There are other different statistical tests, such as t-test, ANOVA, regression, factor analysis and Chi-square. Each has its different set of assumptions and purpose, and choosing one over the other depends on a number of different factors. At this stage, let me mention that I use correlation for exploratory data analysis in machine learning out of these statistical measures. We also use linear and logistic regression in machine learning. However, note that the way we do machine learning is different from how we do it in statistics. I will mention them more in their respective lessons down the line.

Discussing more in-depth these statistics concepts (and how they are calculated) is important but sadly beyond the scope of this book. This book, as mentioned before, is for people who have at least completed their undergraduate curriculum in psychology or other disciplines. For most of the world, the statistics mentioned above are already covered in the curriculum, so you should have gone through those before picking up this book. But if you have not or forgotten a lot of it, I highly recommend you read books on statistics to clarify these concepts. It is not compulsory for machine learning. But it will benefit what we will learn in the upcoming lessons. Alternatively, if you are seeking a shortcut and don't want to read a big fat book on statistics, then search the **bolded** terms such as **confidence intervals** and **effect sizes** on your favourite search engine (e.g. Google, Bing and Yahoo) and read up or watch the videos on these concepts.

References

Aufenanger, T. (2017). *Machine learning to improve experimental design*. FAU discussion papers in economics 16/2017. Retrieved from http://hdl.handle.net/10419/169116

Barranco-GutiÃrrez, A. (2020). Machine learning for brain images classification of two language speakers. *Computational Intelligence and Neuroscience, 2020*, 1–7. https://doi.org/10.1155/2020/9045456

Bolkan, S., & Andersen, P. (2009). Image induction and social influence: Explication and initial tests. *Basic and Applied Social Psychology, 31*(4), 317–324. https://doi.org/10.1080/01973530903316898

Bruhn, M., & McKenzie, D. (2009). In Pursuit of Balance: Randomization in Practice in Development Field Experiments. American Economic Journal: *Applied Economics, 1*(4), 200–232. https://doi.org/10.1257/app.1.4.200

Bzdok, D., Altman, N., & Krzywinski, M. (2018). Statistics versus machine learning. *Nature Methods, 15*(4), 233–234. https://doi.org/10.1038/nmeth.4642

Cheung, M., & Jak, S. (2018). Challenges of big data analyses and applications in psychology. *Zeitschrift Für Psychologie, 226*(4), 209–211. https://doi.org/10.1027/2151-2604/a000348

Cialdini, R., Demaine, L., Sagarin, B., Barrett, D., Rhoads, K., & Winter, P. (2006). Managing social norms for persuasive impact. *Social Influence, 1*(1), 3–15. https://doi.org/10.1080/15534510500181459

Crew, B. (2019). *Google Scholar reveals its most influential papers for 2019.* Retrieved February 26, 2021, from https://www.natureindex.com/news-blog/google-scholar-reveals-most-influential-papers-research-citations-twenty-nineteen

Deng, T., Kanthawala, S., Meng, J., Peng, W., Kononova, A., Hao, Q., et al. (2018). Measuring smartphone usage and task switching with log tracking and self-reports. *Mobile Media & Communication, 7*(1), 3–23. https://doi.org/10.1177/2050157918761491

European Commission. (2019). *AI in Europe: funding hits, 20 billion*. Retrieved February 26, 2021, from https://www.ngi.eu/news/2019/01/08/ai-in-europe-funding-hits-e20-billion/

Freeman, J., & Ambady, N. (2010). MouseTracker: Software for studying real-time mental processing using a computer mouse-tracking method. *Behavior Research Methods, 42*(1), 226–241. https://doi.org/10.3758/brm.42.1.226

Fried, E., & Nesse, R. (2015). Depression sum-scores dont add up: why analyzing specific depression symptoms is essential. *BMC Medicine, 13*(1). https://doi.org/10.1186/s12916-015-0325-4

Geyer, K., Ellis, D., & Piwek, L. (2018). A simple location-tracking app for psychological research. *Behavior Research Methods, 51*(6), 2840–2846. https://doi.org/10.3758/s13428-018-1164-y

Ghosh, C., McVicar, D., Davidson, G., & Shannon, C. (2021). Measuring diagnostic heterogeneity using text-mining of the lived experiences of patients. *BMC Psychiatry, 21*(1). https://doi.org/10.1186/s12888-021-03044-1

Halsey, L. (2019). The reign of the p -value is over: what alternative analyses could we employ to fill the power vacuum?. *Biology Letters, 15*(5), 20190174. https://doi.org/10.1098/rsbl.2019.0174

Hampshire, A., Highfield, R., Parkin, B., & Owen, A. (2012). Fractionating Human Intelligence. *Neuron, 76*(6), 1225–1237. https://doi.org/10.1016/j.neuron.2012.06.022

Harari, G., Lane, N., Wang, R., Crosier, B., Campbell, A., & Gosling, S. (2016). Using smartphones to collect behavioral data in psychological science. *Perspectives on Psychological Science, 11*(6), 838–854. https://doi.org/10.1177/1745691616650285

Johnson, D. R., & Creech, J. C. (1983). Ordinal measures in multiple indicator models: A simulation study of categorization error. *American Sociological Review, 48*, 398–407.

Kunda, Z., Fong, G., Sanitioso, R., & Reber, E. (1993). Directional questions direct self-conceptions. *Journal of Experimental Social Psychology, 29*(1), 63–86. https://doi.org/10.1006/jesp.1993.100

Li, J., & Tong, X. (2020). Statistical hypothesis testing versus machine learning binary classification: Distinctions and guidelines. *Patterns, 1*(7), 100115. https://doi.org/10.1016/j.patter.2020.100115

Lundervold, A., & Lundervold, A. (2019). An overview of deep learning in medical imaging focusing on MRI. *Zeitschrift FÃ¼r Medizinische Physik, 29*(2), 102–127. https://doi.org/10.1016/j.zemedi.2018.11.002

Naab, T., & Sehl, A. (2016). Studies of user-generated content: A systematic review. *Journalism, 18*(10), 1256–1273. https://doi.org/10.1177/1464884916673557

Norman, G. (2010). Likert scales, levels of measurement and the laws of statistics. *Advances in Health Sciences Education, 15*(5), 625–632.

Orrù, G., Monaro, M., Conversano, C., Gemignani, A., & Sartori, G. (2020). Machine Learning in Psychometrics and Psychological Research. *Frontiers In Psychology, 10*. https://doi.org/10.3389/fpsyg.2019.02970

Peirce, J., Gray, J., Simpson, S., MacAskill, M., Hăchenberger, R., Sogo, H., et al. (2019). PsychoPy2: Experiments in behavior made easy. *Behavior Research Methods, 51*(1), 195–203. https://doi.org/10.3758/s13428-018-01193-y

Strohmetz, D., Rind, B., Fisher, R., & Lynn, M. (2002). Sweetening the till: The use of candy to increase restaurant tipping1. *Journal of Applied Social Psychology, 32*(2), 300–309. https://doi.org/10.1111/j.1559-1816.2002.tb00216.x#

Sullivan, G., & Artino, A. R., Jr. (2013). Analyzing and interpreting data from likert-type scales. *Journal of Graduate Medical Education., 5*(4), 541–542.

Vanman, E., & Philipp, M. (2019). Physiological measures. In J. Edlund & A. Nichols (Eds.), *Advanced research methods for the social and behavioural sciences* (pp. 149–162). Cambridge University Press.

Whiting, K. (2020). 5 charts showing the jobs of a post-pandemic future – and the skills you need to get them. World Economic Forum. Retrieved 21 August 2022, from https://www.weforum.org/agenda/2020/10/x-charts-showing-the-jobs-ofa-post-pandemic-future-and-the-skills-you-need-to-get-them/

Zumbo, B. D., & Zimmerman, D. W. (1993). Is the selection of statistical methods governed by level of measurement? *Canadian Psychology, 34*, 390–400.

Chapter 2
Python Programming

Learning Outcome of This Chapter After reading this chapter of the book, students will be able to *read and understand the codes to effectively use the Python codes for data analysis for their research projects.*

Communication is the key to all relationships, whether it is with your partner, parent, children or computer. Yes, computer – the toolbox with which people like you and me share a long-term relationship. We probably spend more time with a computer than our spouses and children. We interact with computers for various purposes in our everyday life, from checking our social media accounts, writing a word document, watching movies, or shopping for a dress from an e-commerce store. Every time we do so, whether we realise it or not, we are communicating with the computer. We are telling it to do something for us. It can be a simple mouse-clicking activity on icons or typing the alphanumeric keys on the keyboard. But then there are times when we need to communicate with the computer more directly using the language it understands. One such time is when you need to ask the computer to perform some complex computations for you.

Just as we humans have different languages such as English, German, Bengali and French, computers also have languages such as Python, Java, C++ and R. These are called programming languages. In this part of the book, we will learn Python language so that we can use it later to tell the computer to analyse the dataset.

Our choice will be Python. The reason is simple. It's the most popular programming language in the world right now. You will learn to do data analysis with Python, but you can also use Python to build web apps, games, design experiments and a lot more if you learn it well. For these reasons, Python programmers get paid a lot of money. Plus, there is high demand for Python coding services. So, even if someday you wish to retire from psychology, and if you know Python well enough, then you will be easily able to find a job or at least provide your coding services through freelancing marketplaces.

There are fans of R but hey, Python rules. Trust me.

C. Ghosh, *Data Analysis with Machine Learning for Psychologists*,
https://doi.org/10.1007/978-3-031-14634-3_2

2.1 But Do I Have to Learn to Code for Data Analysis?

For the long term: Yes. There is open-source data-analytic software such as KNIME (https://www.knime.com/downloads). You can perform data analysis without any kind of programming or Python codes. But there are limitations and restrictions to what and how much you can do with it. Learning Python liberates you into endless possibilities. Plus, it's good to have Python on your resume, isn't it?

Bonus
The statistical software platform we learn at our University might not pull the load of such big data, and a personal computer is likely to hang up or crash. This happened to me. Like any other psychology student, I was used to some of the most popular statistics software in the market to analyse the data. They worked well for small data sizes. But during my PhD studies, when I used a dataset of 11,000 patients, it miserably failed to process it because the software kept crashing. Initially, I thought it was my old computer. But when I used the same dataset on Python, it worked smoothly as a breeze. The point is that when you are processing a large dataset, such software packages might fail, and you might need to use Python or R programming languages. That said, if you have super-big data, you need to get yourself a high-end computer and get Python. Evaluating a large dataset with Python can consume a lot of memory and/or disk space. One of my computer science friends regularly runs simulations that use 32GB of memory and hundreds of GB of disk space. But as a researcher in the school of psychology, the amount of data I have dealt with so far worked reasonably well with my standard computer. So, I imagine you are safe with a standard computer with Python installed on it.

Suppose you are experiencing self-doubts or just a general aversion to programming. In that case, that is a natural feeling for a few people. So just hold on with me till the end of this chapter. I promise I will make Python fun and easy for you. If it does not, you don't have to progress to the next Chapters of this book. Deal?

Let's get started.

2.2 How to Install Python?

We will learn Python 3 in this book. It is the latest version. Python 3 is the most current release, and that version 2 still exists but is not really supported any longer. Suppose you install the older version by accident. In that case, many examples presented in this book won't work because the syntax is different, and some libraries won't work.

Python is free of cost. In fact, all of the tools you will learn in this book is open source ("free").

You cannot write and run Python codes in your Word document or Excel. You will need a special type of platform to write codes. There are several. But for starters, I recommend Spyder or Jupyter.

Let's stick to Jupyter (https://jupyter.org/) in this book. You can download it from the Anaconda Python distribution platform (https://www.anaconda.com/products/individual) or its original website. I recommend installing it via Anaconda because it is easier to navigate through files in Jupyter and install new libraries, packages or modules (which are important, as you will see later).

Installing Anaconda is pretty straightforward and, of course, free of cost. Just search for the link and download the relevant version (Fig. 2.1).

Note that there are different versions of Python. A çomplete list can be found with the release dates on the Python official website (https://www.Python.org/downloads/). A new version of Python gets released every other month. But do not worry. Luckily, Python 3.x versions are all very similar. There are no significant differences in the syntax. So, whichever version you pick, if it is Python 3.x, it should not be a problem for you. In this book, we shall learn 3.8.1 (the stable version used when writing this book).

Once you complete the installation process, you can launch Jupyter directly from search into your system (Fig. 2.2).

After opening, this is the first screen you will see (Fig. 2.3).

Click on New>Python 3

It would open on a screen such as this. In Jupyter, the word "cell" is a multiline text input field. Its contents can be executed by clicking on RUN in the menu bar or by using shift-enter on keyboard (Fig. 2.4).

If you click on file, you will see the options in Fig. 2.5:

The "download as" option lets you download the Jupyter script into a .py file which can then be run on a different computer (which probably does not use Jupyter but some other coding environment).

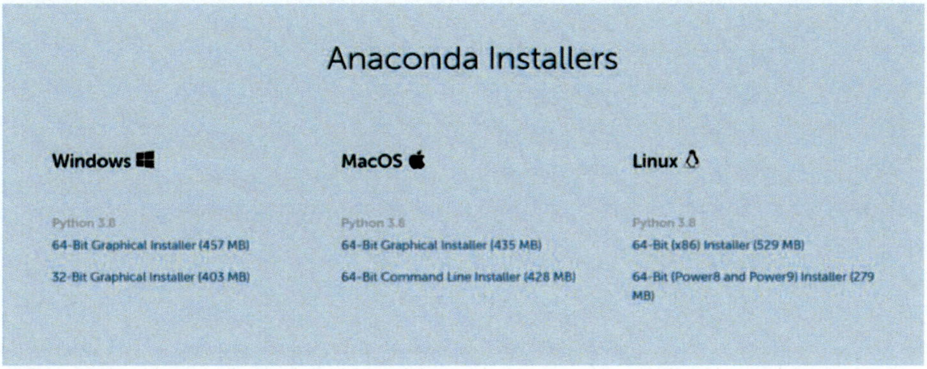

Fig. 2.1 The different installation options (subject to update frequently)

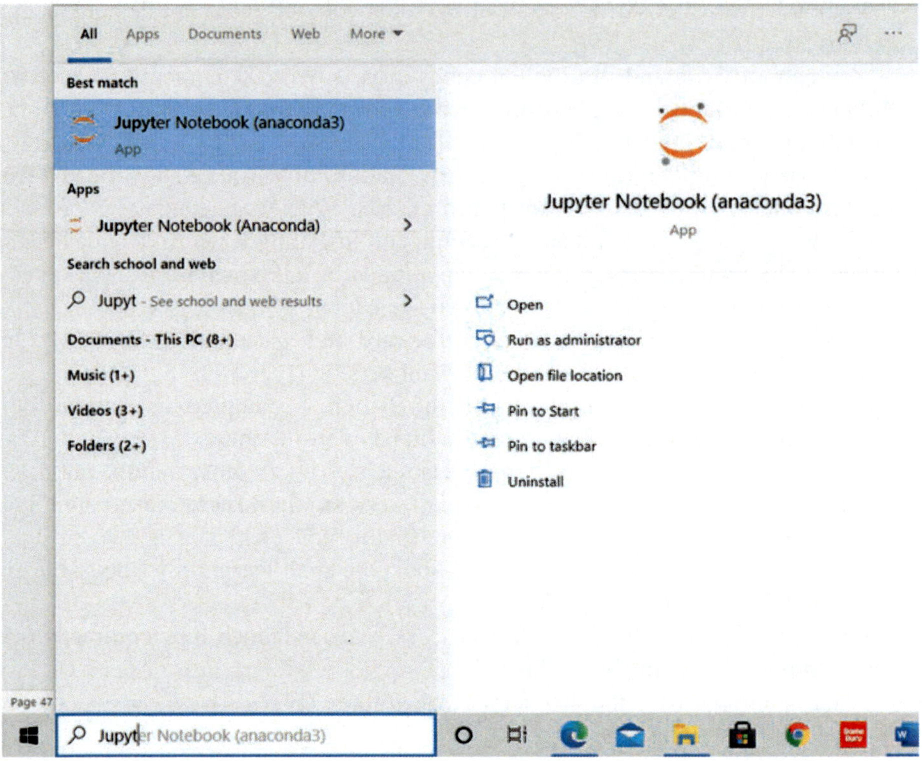

Fig. 2.2 The Jupyter Notebook icon when searched from the search option of Microsoft Windows

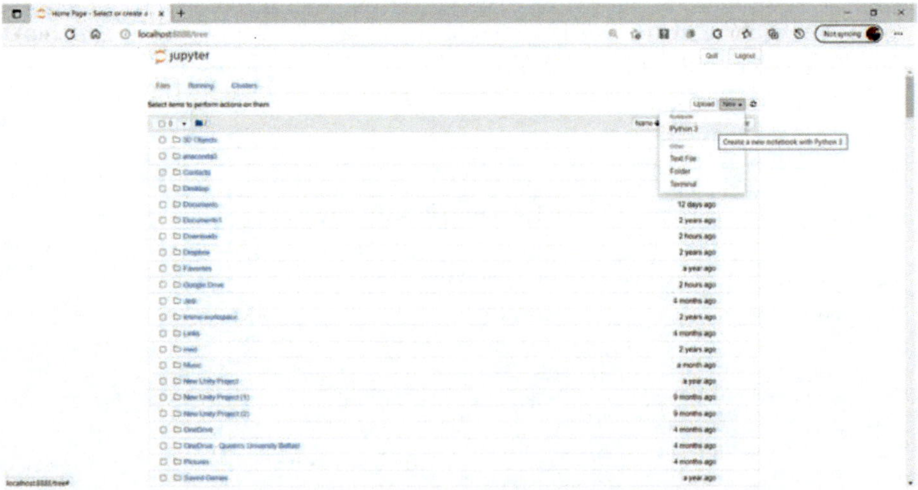

Fig. 2.3 The opening screen of Jupyter showing all the files from the past and options to create a new Python 3 file

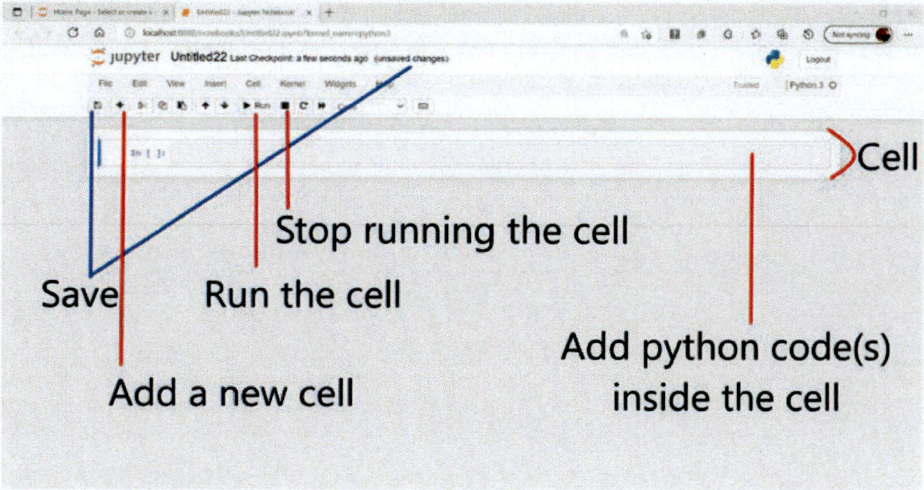

Fig. 2.4 The view of Jupyter window

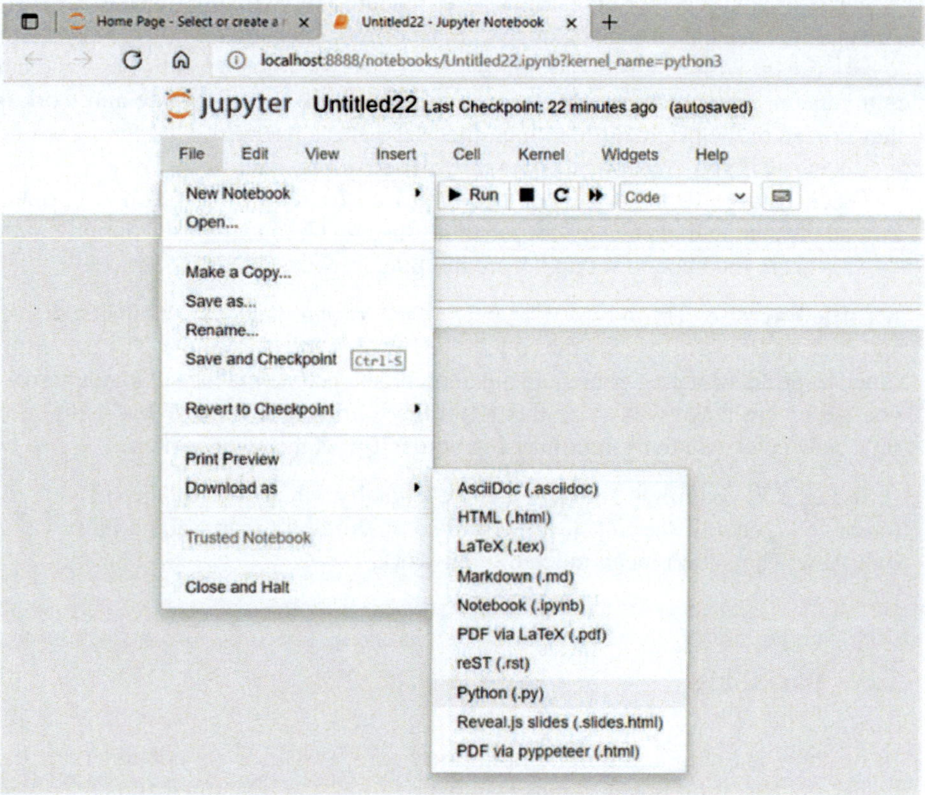

Fig. 2.5 Different extension options to download the code script from Jupyter

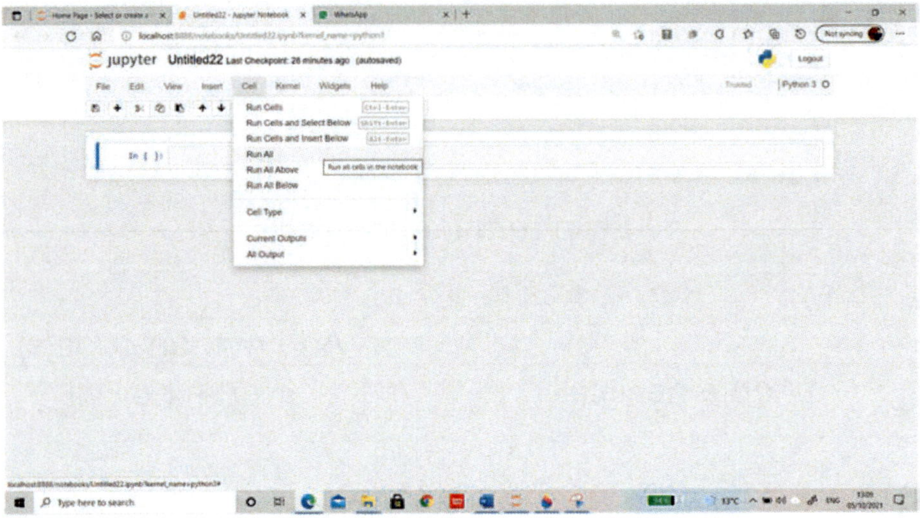

Fig. 2.6 Different options to run the cell in Jupyter

Next, if you scan through the list of menus, you will see "Make a Copy". Personally, I use the "Make a Copy…" most often, especially at times when I need to experiment with a substantial number of codes, but I don't want to lose the current version. In such cases, it's easier to create a duplicate of the file and work on that instead of the original file (which then serves as a backup).

Likewise, if you click on "Cell", you will see in Fig. 2.6:

When I have 10 or 20 cells of codes to run, I click on the "Run All" option instead of running them one by one, so that's the most frequently used option for me.

This is all you need to set up before jumping to codes.

A Little Exercise Try to type, `print ("My Name is XYZ")` inside the cell and click on RUN (to execute the code). See Fig. 2.7. for an example. The `print()` function prints whatever contents you entered between its brackets. Do not worry if you get an error. We will cover this all in the following lessons. What's important right now is for you to try familiarising yourself with the environment.

In Fig. 2.7, you might see In [1], and eventually, when you run the cell or a different cell, you will see In [2], In [3] and so on. So don't worry about those values right now. They don't mean much to your work.

2.3 Variables

In machine learning and programming literature, variables are containers of data values (i.e. variables hold references to data). So, if `X = 10`, then 10 is the object, and X is the variable here. Likewise, objects can also be listed such as [5, 10, 20, 40]

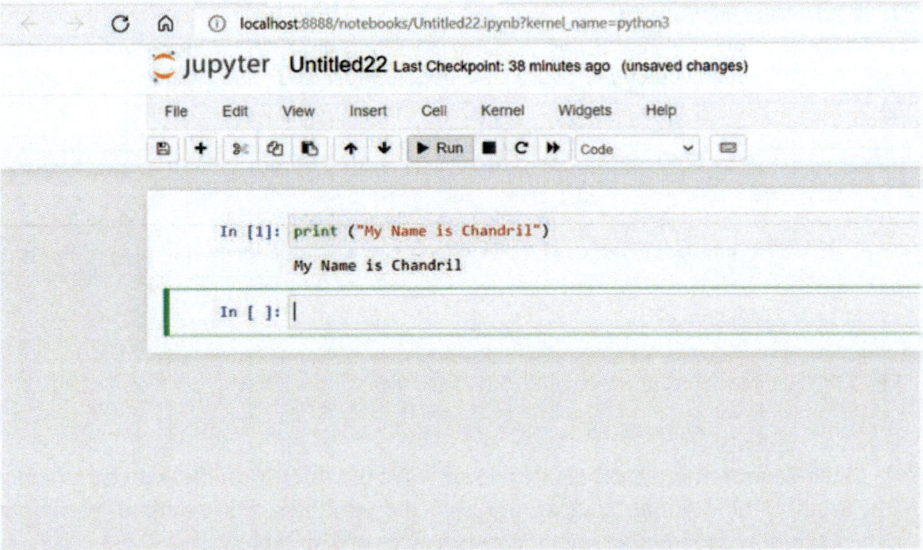

Fig. 2.7 An example of Python code

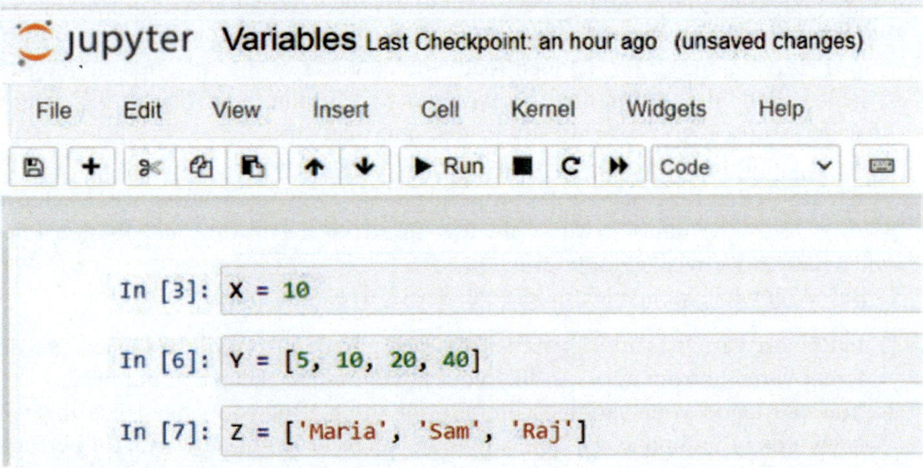

Fig. 2.8 Examples of variables in Python

and a variable such as Y can be assigned to it (i.e. `Y = [5, 10, 20, 40]`). Variables can also store texts (not just numbers). An example might be `Z = ['Maria', 'Sam', 'Raj']`. Note that variables come in different types. For example, `X=10` is a typical single-value variable, but `Z = ['Maria', 'Raj', 'Sam']` is a specific Python variable type called a list.

On Jupyter Notebook, it will look something like as in Fig. 2.8.

Table 2.1 An example dataset

Participant name	Age	Low mood	Stress	Hearing voices
Maria	25	0	1	0
Sam	32	1	1	0
Raj	19	0	1	1

```
In [ ]: df = pandas.read_csv('D:\\dataset.csv', header = [0])

In [ ]: df.describe()
```

Fig. 2.9 Example of creating a variable ("`df`") and then calling it later ("`df.describe()`")

In the dataset, there can be a variety of ways in which variables can be assigned. But usually, the column headers represent the variables. For example, participant name, age, low mood, stress and hearing voices are in Table 2.1.

In the code below in Fig. 2.9, "`df`" is the variable assigned to the line of code that is reading your dataset. It would be tedious and, at times, troublesome if you have to type "`pandas.read_csv('D:\\dataset.csv', header = [0])`" every single time you wish the code script to use the dataset to perform your desired function or task. You can assign it to a variable such as df to avoid that. The code `df = pandas.read_csv(...)` then loads the contents of the specified file into the variable called "`df`". From there, you reference it for later use. Using "`df`" later in the code just uses what was already loaded into that variable.

For example, in the codes below in Fig. 2.9, we assigned `df` to `pandas.read_csv('D:\\dataset.csv', header = [0])`, so when we wanted to "`describe`" the dataset, we could use `df.describe()`, and Python knew which dataset we wished to describe.

Based on what we just learnt, we can derive five conventions of Python:

1. In Python, variables are case-sensitive, meaning that *Age* will be treated as a different variable from *age*. Just like your password. So, keep that in mind!
2. You can name your variables literally anything. But you cannot use reserved keywords in Python as variables. Python has reserved these words for performing certain specific tasks and won't allow you to use them as variable names. The reserved keywords are as follows: `False, def, if, raise None, del, import, return True, elif, in, try and, else, is, while as, except, lambda, with assert, finally, nonlocal, yield break, for, not class, from, or continue, global, pass.`
3. In mathematics, `x = 3` means x is equal to 3. But for Python `x = 3` means the variable x currently stores 3 as its value. If you want to tell Python that x equals 3, you have to write it as `x == 3` (use double "equal to-" signs).
4. You cannot put blank spaces between two words of a multi-worded variable. For example, if your variable name is "low mood", then you must enter it as "low_

Fig. 2.10 An example code to print an output in Python

```
In [11]: X = 10

In [12]: print (X)

         10
```

mood" or maybe as "LowMood", but you cannot leave blank spaces in between these two words.

5. While assigning text-based data (also known as "strings" in Python), you need to put them within quotation marks (single or double). For example, see above where we assigned `Z = ['Maria', 'Sam', 'Raj']`. You can enter `Z = [Maria, Sam, Raj]`. The quotation marks are important for assigning text data.

How to Print a Message or Result in Python?
We are writing codes to generate an output such as a statistical result. Right?

We use the print function for that. The simplest example would probably be as in Fig. 2.10.

Note A dataset can have features and dimensions. Variables are not generally equivalent to features but can depend on the code's intent. For a beginner's heuristic, take those variables, features and dimensions are synonymous. Still, as said in the previous sentence, there might be times when they are not. In this book, we will simplify it by displaying examples where they are all the same (which will be for most of a psychologist's needs). Therefore, when you come across concepts such as feature importance and dimensionality reduction, you can think them of as variable importance and variable reduction (for now).

2.4 Operators

2.4.1 Arithmetic Operators

Some time back, we saw how we can assign data to a variable using an equal to sign (=), an operator. More specifically, it would be an example of an assignment operator. Then we also came across the following example of an operator (+) that enabled us to perform basic arithmetic as in Fig. 2.11.

If you noticed, we used a summation sign between two variables to add them up. Such operators (e.g. +, −, *, /) are called arithmetic operators and are the same as the ones we use in mathematics.

In addition to "addition", you can also perform other arithmetic operations. Some examples are demonstrated in Fig. 2.12.

```
In [13]: X = 10

In [14]: Y = 2

In [15]: print (X+Y)

12
```

Fig. 2.11 Example of printing a simple arithmetic output

```
In [13]: X = 10

In [14]: Y = 2

In [15]: print (X + Y)

12

In [16]: print (X - Y) #subtraction

8

In [17]: print (X * Y) #multiplication

20

In [18]: print (X / Y) #division

5.0

In [19]: print (X ** Y) #Power  (a raised to power b)

100
```

Fig. 2.12 Example of Python statements to perform basic arithmetic operations

In addition to *arithmetic operators* (e.g. +) and *assignment operators* (e.g. =), there are several other types of operators in Python such as *logical operators* (e.g. and, or, not), *identity operators* (e.g. is, is not), *membership operators* (in, not in) and *bitwise operators* (e.g. &, |, ~ representing AND, OR and NOT respectively). But more often, I have come across *comparison operators*. Since this book is meant to help you get started, I will talk about *comparison operators*. I shall leave the other operators for you to explore once you complete this book.

Let's learn about *comparison operators* in this book. Together with arithmetic operators, they should cover most of a beginner's needs.

2.4.2 Comparison Operators

Comparison operators, put in simple terms, are used to compare two values, such as greater than, lesser than, greater than equal to and so forth – same as in mathematics. Examples of such comparison operators are as follows:

- Not Equals: `x != y`
- Less than: `x < y`
- Less than or equal to: `x <= y`
- Greater than: `x > y`
- Greater than or equal to: `x >= y`

But where do we implement these comparison operators in Python?

We can use such operators to set *conditions* in Python. So what are the conditions?

In life, we grew up with hearing conditions. As a child, you might have been told that "If you do your homework, then you can go out and play…", "If you be a good kid at the shopping mall, you will get a chocolate...else you won't get it" and so forth. These if-then statements are conditions. In our daily life, we employ conditions as well. For example, IF it's raining, THEN carry an umbrella. Else, don't carry an umbrella. In Python, we can do the same. For example, you can tell Python, `if a>b`, then `print ("a is greater than b")`.

Take, for example, IQ scores. Suppose you are creating a simple computer program to tell users whether the person has superior intelligence. See the codes in Fig. 2.13. In the first Jupyter cell, the user enters the IQ test scores. The second cell (block of codes) compares the input value with the pre-fixed criteria (if the IQ score is greater than equal to 120) and prints an output accordingly.

Take a note of the code in Fig. 2.13 and notice the indentations. Indents are required for conditional blocks of code. But in general, an indentation in Python is very important for it represents a block of code. The colon you notice is used to represent an indented block. If you don't give proper indentations, Python will throw up an error (something like "`IndentationError: expected an indented block`"). But it's not difficult to keep in mind. For example, in Jupyter, when you HIT enter key after the colon punctuation (:), it automatically goes to the next line and starts with an indent.

```
In [20]: IQ_Score = 125

In [21]: if IQ_Score >= 120:
             print("This person has superior intelligence")
         else:
             print("This person does not have superior intelligence")

         This person has superior intelligence
```

Fig. 2.13 Example of Python codes to demonstrate IF-ELSE statements

```
In [22]: IQ = 45
```

```
In [30]: if IQ > 120:
             print("This person has superior intelligence")
         else:
             print("This person does not have superior intelligence")

         This person does not have superior intelligence
```

```
In [26]: if IQ >= 120:
             print("This person has superior intelligence")
         elif IQ >= 95:
             print("This person has an average IQ")
         else:
             print("This person has below average IQ")

         This person has below average IQ
```

Fig. 2.14 Sample of Python codes to demonstrate if, else and elif statements

2.5 Statements

In the previous lesson, we got introduced to *if-else statements*. Now let us see them in action in Fig. 2.14. Additionally, in this lesson, we will learn about `elif` and `if...else` statements. `Elif` tells Python, "if the previous conditions were not true, then perform this condition". Likewise, if...else statement tells Python to do SOMETHING if a particular condition is met, else do SOMETHING ELSE.

Accordingly, the three statements are:

- if statement
- if-then-else statement
- if and elif statement

Before you go: Did you notice an indentation before print?

2.6 Loops

Imagine you need to print a message "Thank you for participating in our study" using Python. How will you do it?

Based on what you learnt so far, you would probably arrive at a solution as shown in Fig. 2.15.

This is correct. But suppose the task is to print "Thank you for participating in our study" a whopping 100 times. Taking this approach, you need to type `print()` function 1000 times. That sounds like a lot of effort. Luckily there is a smarter way to do this. You can use loops to automate such boring tasks. There are two types of loops: definite loops and indefinite loops.

```
In [1]: message = "Thank you for participating in our study"

In [2]: print (message)

        Thank you for participating in our study
```

Fig. 2.15 Simple Python code for printing a message

```
In [3]: for message in range(0, 1000):
            print("Thank you for participating in our study")

        Thank you for participating in our study
        Thank you for participating in our study
        Thank you for participating in our study
        Thank you for participating in our study
        Thank you for participating in our study
        Thank you for participating in our study
        Thank you for participating in our study
        Thank you for participating in our study
        Thank you for participating in our study
        Thank you for participating in our study
        Thank you for participating in our study
        Thank you for participating in our study
        Thank you for participating in our study
        Thank you for participating in our study
        Thank you for participating in our study
        Thank you for participating in our study
        Thank you for participating in our study
        Thank you for participating in our study
        Thank you for participating in our study
        Thank you for participating in our study
```

Fig. 2.16 Example code for creating "definite" loops in Python

Definite Loops Are FOR Loops Simply put, FOR loops enable us to iterate over a sequence (automate the boring task of printing 1000 times). The best part is it's just two new lines of code.

It will look something, as shown in Fig. 2.16.

In the codes above, we used FOR loop for iterating over a single message. Of course, if you want, you can also iterate over a sequence (such as a list or string). But for the sake of simplicity, we focused on this exercise of printing a message.

Indefinite Loops Are WHILE Loops

Did you notice with just two lines of code, we can now print the same message as many times we wish. How cool is that?

But here's a catch: For using a FOR loop, you need to be absolutely sure how many times you wish to print the message. In other words, you need to know how many times you need to `PRINT` the message. What if, in your use case, you don't know the exact figure? It can be 900, but it can also be 1100. We will use a WHILE loop in such a case. You can use the FOR loop when you know a specific length. In

```
In [8]: while True :
            message = input("Complete the survey Yes/No:")
            if message in ['Yes']:
                print("Thank you for participating in our study")
                break
            elif message in ['No']:
                print("Sorry to see you go")
                break

        Complete the survey Yes/No:Yes
        Thank you for participating in our study
```

Fig. 2.17 Example code for creating an "indefinite" loop in Python

this case, we knew in advance that we needed to print the message 1000 times. When we don't know how many times we need to perform a particular task (in this case, it is "printing the message") in advance, we will opt for the WHILE loop, which will keep performing the task ("printing the message") until a given condition is met. This condition can be something like the room temperature or time. An example task could be keeping the radiators switched on until the room is 18 °C.

Imagine a use case where the computer displays "Complete the survey, Yes or NO" – it requires the user to respond with Yes or No. If the user enters Yes, you want the computer to show "Thank you for participating in our study", and if the user enters No, then the computer must display "Sorry to see you go". The codes might look as in Fig. 2.17:

The idea is simple. The While loop works by iterating over a block of code as long as a specified condition is true. In this case, as long as it gets an input "Yes" or "No", it will keep generating the message.

Just to give you a sense of how the While loop works, here's another real-world example: Imagine an automated baggage security scanner at the airport. It will allow all the bags to pass through till a restricted item (e.g. gun or bomb) is found. When a bag enters the scanner, it is checked whether any weapons (or other restricted items) are in the bag. Let the test expression, in this case, be "safe_items" = True. After one iteration, the test expression is rechecked. This process continues until the scanner detects the presence of a dangerous item ("safe_items" = False).

A simpler example of a While loop might be as presented in Fig. 2.18:

In this case, the user input is participant attendance = 6. Since the while is >0, it'll run until the variable value reaches 0. We have introduced a modifier (`Participant_Attendance -= 1`), which is compulsory in this case because if you don't decrement the counter, it will never change. Try running that code block without changing the value, and watch the endless loop.

Once the code is run with 6 as an input (example), it first checks if the condition (greater than 0) is met. If it does, then the loop is run in iteration with each time the input is reduced by 1 (modifier, -=1). Thus, the code will be executed in a loop until the time the condition is broken (when six is reduced to 0). If you still struggle with what an "iteration" is, then it's simply referring to each repetition of the process.

```
Participant_Attendance = 6

while Participant_Attendance > 0: #The controlling expression is initialised before we start the loop.
                                  #It is modified later in loop.
    Participant_Attendance -= 1   #Participant_Attendance = Participant_Attendance - 1.
                                  #The expression is modified here (decremented by 1).
    print(Participant_Attendance)
```

```
5
4
3
2
1
0
```

Fig. 2.18 A simple example of a “while” loop in Python

One repetition is a single iteration, and each repetition’s result is the starting point of the next iteration.

The code we wrote using the While loop checks for the condition at a particular instance , whether the number of participants is still more than zero. While the conditions are true, the lines of code below are executed, subtracting the attendance by 1 and then printing them. This happens in a loop. The loop runs every time till the attendance condition is no longer true. When the participant attendance reached the “0”, the condition Participant_Attendance > 0 became FALSE. Then, the iteration stopped.

P.S. I am sure not many of you might be interested in building a computer program that will show your participants a “Thank you” message. I won’t be too. But I am using that as a concrete example here for a better grasp of the concept. Later you will observe how we use the FOR and While loops to analyse data.

Before you go: Did you notice the hashtags in the code?

Putting a hashtag makes everything after it into a `*comment*`. Putting triple quotations (`"""` or `'''`) also before and after a part of text or code also tells Python to ignore that portion while executing the code script. It is useful for short comments or reminders for future reference. Remember, while you are using `"""` or `'''` for comments, previously, you have used them to delineate strings (e.g. `Message = "Hello, world."`).

2.7 Data Structure

We will be learning a lot about analysing data in this book. But before, let us take a quick familiarisation with some basic terms and concepts related to data in Python.

Types of Data

Python can process any type of data: words, numbers, images, audio and even video. But in this book, we are just getting started and so will keep ourselves restricted to learning to analyse numbers. In this context, you need to know four types of data here in the basics of Python coding.

- Text data (which are called "**string**")
- Numeric data (two types include **integers** (without decimal) and **floats** (with decimal))
- Sequence of data (**list**, **tuple** and **range**)
- Boolean (**true** or **false**)

On Storing Collections of Data as Strings, Lists, Tuples or Dictionaries

1. List (e.g. `Xlist = ["memory", "attention", "creativity"]`). Allows future modifications (i.e. change, add or remove items after it has been created) and allows duplicate values.
2. Tuple (e.g. `Xlist = ("memory", "attention", "creativity")`). Unchangeable but still allows duplicate values.
3. Set (e.g. `Xlist = {"memory", "attention", "creativity"}`). Cannot be modified and do not allow duplicate values.
4. Dictionary (e.g. `Xlist = {"name": "Martha", "age": 54, "Year": 2021}`). Allows future modifications but does not allow duplicate items with the same key. Little exercise: Try `Xlist = {"name": "Martha", "age": 54, "age": 13, "Year": 2021}` on Spyder. You will probably see that the code shows an error. A tiny yellow triangle with an exclamation mark will appear at the start of the code line. Hover the cursor over it. Notice what it suggests about "age".

Note We will use only list and tuple in this book. Tuples are more memory and time-efficient than the lists. So, if your data is not meant to be changed in the first place, you should use tuple data type over lists.

Remember, my aim here is to familiarise you with the terms and jargon frequently used in Python and machine learning. So, it is OK if you don't quite understand these yet. We will learn them all later when we integrate entire code scripts.

2.8 Methods and Functions (Built-Ins) in Python

Methods and functions enable us to perform certain tasks in Python. I cannot imagine doing data analysis without them. So, let me put some special emphasis on these right at this stage.

2.8.1 Methods

Methods are used to alter an object's state. So, if the object is "Thank you for participating in our study", you can use a method like `.upper()` to change it into upper case. See the code in Fig. 2.19, for example.

Now, how do you know which other methods are available for use?

It's simple. In Jupyter Notebook, after you type in your variable, put a dot (full-stop punctuation) and then hit the TAB key. You will see all the possible options popping up. See the code in Fig. 2.20, for example.

Now While Reading Someone Else's Python Script, How Will You Identify Which Element Is a Method?

Simple. Methods come after a dot attached to a variable/object. For example, if you remember, in the part where we spoke of "Variables and Operators", we used `df = pandas.read_csv('D:\\dataset.csv', header = [0])`. In here, `.read_csv()` was the method. Later in this book, when you read my codes for machine learning, you will spot plenty of these methods. Few examples would be `.append()`, `.KFold()` and `.figure()`. You will learn about them later. To sum up, methods are called on an object or variables representing an object. You can spot them easily as they come after a dot in Python scripts.

Although a useful heuristic for absolute beginners, if you really care about what it is called, then there are exceptions. The function `head()` comes after a dot, that is, dependent on an object. So, it is written as `data.head()` which means to show the top five rows of the dataset (five is the default, you can change it by placing a number within the parentheses). In that case, `.head()` is a function.

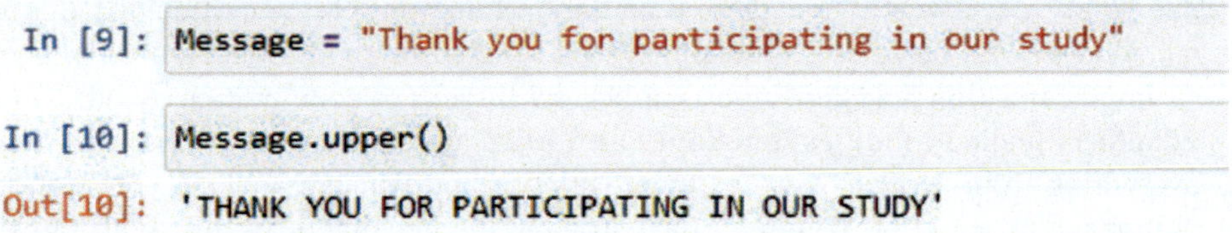

Fig. 2.19 Example of using method (".upper()") in Python

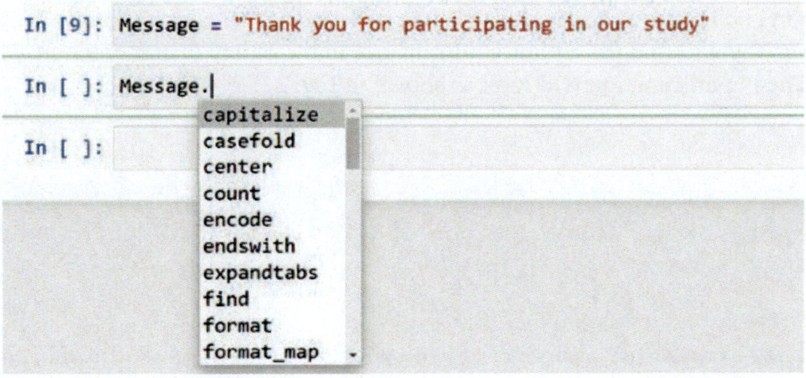

Fig. 2.20 Available "Methods" for use

2.8.2 Function

Unlike methods which is called upon an object, functions are standalone. For example, in the code presented in Fig. 2.21, we called the function "`len ()`" against the variable x, it's printing 40, which indicates there are 40 characters (includes blank spaces) within this string ("Thank you for participating in our study").

Other In-Built Functions in Python Are as Follows

`print(), abs(), delattr(), hash(), memoryview(), set(), all(), dict(), help(), min(), setattr(), any(), dir(), hex(), next(), slice(), ascii(), divmod(), id(), object(), sorted(), bin(), enumerate(), input(), oct(), staticmethod(), bool(), eval(), int(), open(), str(), breakpoint(), exec(), isinstance(), ord(), sum(), bytearray(), filter(), issubclass(), pow(), super(), bytes(), float(), iter(), tuple(), callable(), format(), len(), property(), type(), chr(), frozenset(), list(), range(), vars(), classmethod(), getattr(), locals(), repr(), zip(), compile(), globals(), map(), reversed(), import(), complex(), hasattr(), max(), round().`

If you notice the first in-built function on the list, it is the print function (I have underlined it for your convenience). Yes – the same one we had been using all along this time. Now, if you notice, there is an import function. This is particularly important, and we will be using it frequently, so it is better to be introduced now.

Import function helps us import (or borrow) blocks of code written by other programmers (called **modules, packages and libraries**). So, you don't have to write everything from scratch. For example, suppose you want to analyse your dataset with machine learning. You don't have to write the machine learning algorithms in that case. There are plenty of those available open sources (free!!!), which you can import into your code and make it run on your dataset and generate results. We will learn how to do these later in this book, but note that the import function is a lifesaver. You possibly cannot do data analysis so seamlessly without this ability to import and borrow others' code. It's free of cost, but make sure you cite them in the paper. You will find the author's information in the online documentation.

The import function will look as shown in Fig. 2.22:

```
In [11]: x = "Thank you for participating in our study"

         len (x)

Out[11]: 40
```

Fig. 2.21 Example of using function ("len()") in Python

```
In [185]: import pandas as pd

In [186]: import numpy as np
```

Fig. 2.22 Example of using the "import" function

Before You Go You can create new methods and functions of your own. There are also topics such as **classes** and **objects**. However, in my experience, it is rarely needed in a beginner's practice. Instead, they are more helpful for advanced Python and data science pursuers. In each lesson of this book, I have reminded myself that I am writing this book for absolute beginners – people who come from a social science or health science disciplines but without any background in computer science or programming. In my experience with teaching students at the university, I have felt that starters are typically shy and scared people. They need a smooth and wish for a quick start. They want to learn as much as is required to get their job done without getting into the needless complexity of the subject matter. Therefore, I have not included how to write your own methods and functions in this book because I believe it is not needed right now. Once you gather confidence and love for the subject, you will pick up an advanced book on the subject to learn such topics.

2.9 Error Resolution

Getting error messages is part and parcel of coders' daily life. So do not worry about that. Even the experienced ones rarely get their codes run without any errors for the first few times. Fortunately, there are easy ways to deal with them. Here's how:

#1 Path: Python and machine learning have large online communities (social media). You can join Facebook groups on these topics and ask your queries in the group. Someone or the other will respond to you. You will be surprised to see so many people on the Internet will be willing to help you more than your closest friends ever did. Once, a senior academic from Harvard University spent 8 hours straight with me to resolve an error. I never met this person, and he was not getting any money from me either. He was just a kind and nice person. There are many such good people there. All you need to do is just ask for help.

#2 Path: Likewise, there are websites such as Stack Overflow (https://stackoverflow.com/). You can post your questions to get help from random strangers. You can return their favour when you learn it well by helping the next generation of coders and machine learning students altruistically. That's how it works.

#3 Path: But you do not have to post your questions or error message whenever you are stuck. Because it's not the most time-efficient approach to get help, people might take days even to notice your post and then respond to it. Honestly, there have been plenty of times when I did not respond to my posts on Stack Overflow. So, if you have a deadline and don't have much time, then there's something else you can do. You can copy the error message and paste it onto your favourite search engine on the web and search for similar search queries. The chances are that someone has already encountered the same or at least a similar error message in the past. That person has posted the question online on some website and has received feedback or help from experts. You can read their proposed solutions and try them on your code script. I use this path of seeking help 99% of the time. It's easy and quick.

Look below for an example in Fig. 2.23:

2.10 Last Words

That is all about Python programming for now. If you have not already done so, I recommend you go back and re-read the eight lessons and do this on your own computer. Even if it looks very simple, it might be tempting to go to the next step. Still, I strongly recommend you to imitate whatever I have taught and see if you are getting the same results. There might be times when you get an error; that's when you will realise it was not as simple as it seemed. If you get stuck with an error, as said before, read the error message. If that does not make sense, copy-paste it on your web browser to check what solutions were offered to people who had previously raised similar questions. Hopefully, you will not get errors at this stage, and it should be a smooth ride. Now, as promised: did it seem easy? If not, then you don't have to move to the book's next chapter. But just in case you feel that I kept my promise, I am excited to see you in the next chapter, where we will utilise Python to refine the raw data.

Bonus

Much of the developments on the lines of Python and machine learning happened in the United States. Probably for that reason, you will find American English in most places. I have used British English all my life. But Python only understands American English. I will give you an example. You will find words like visuali-"z"-er and not visuali-"s"-er. Since these are conventions of the coding world, we cannot change it to visualiser in the script, especially when importing or using a Python library. But while assigning variables, you can do anything. Therefore, if you ever find me using -"s"er and not -"z"er - tell yourself that you can write anything there, and it won't matter. I might type import visualizer because I am calling a library. Still, I can write `visualiser_prediction_error = PredictionError(RandomForestModel)` because here visualiser is a variable name. You find will an example on page 109 under the topic of regression.

Fig. 2.23 Screenshot of Stack Overflow for an example of Q/A help

To end, remember that Python is a general-purpose programming language. Its scopes are huge. You can make web applications, computer games and a lot of other useful things with Python. But you will probably not do or need to do any of that. You are a psychologist wanting to learn Python to get started with data analytics for your research project. For this purpose, the eight lessons you learnt before are more than what you will probably need in daily life. But just in case you want to do other kinds of work, such as build software or maybe transition from being a psychologist to a full-time Python programmer, you will probably need to delve a lot deeper. In this book, we only scratched the surface of Python coding – just as much you need to get started. So sadly, you are not ready to apply for the post of Python developer yet. But the good news is that now you know Python as much as you need to complete this book successfully.

Chapter 3
Data Pre-processing

Learning Outcome After reading this chapter of the book, students will be able to *prepare the raw ("dirty") data for data analysis* to *build better machine learning models*.

3.1 Introduction

In the first chapter, we learnt about the different sources of data. Imagine a survey design. After administering the questionnaires to probably 1000s of people, you or your research assistant have entered the data into the spreadsheet. However, it is not before you glimpse through the entire dataset on a spreadsheet that you will realise that the dataset is not ready for analytics yet. There are four potential major issues with raw numerical datasets:

1. Some people might not respond to all the items on the questionnaire (or scale). In other words, you have "missing data".
2. Likewise, the person who entered the data might enter the wrong values (occasionally). For example, instead of entering age 30, you might have entered it as 300 (with an extra zero).
3. The data you collected has numerous variables on different scales. Some such as age is between 18 and 60 years, but others such as salary are £20k and beyond. While the difference in numeric values is obvious to a human eye, but for the dumb computer it does not understand the difference between units of measurements (i.e. years of age vs currency of money) will weigh bigger values higher. Therefore, salary and age will not be treated equally.
4. There might be a lot of variables (probably hundreds) you collected to predict variable Y. But not all of the independent variables contribute equally to the

C. Ghosh, *Data Analysis with Machine Learning for Psychologists*,
https://doi.org/10.1007/978-3-031-14634-3_3

dependent variable. So, the model will have many needless independent variables (which probably contribute ZERO to the target-dependent variable).

There can be additional issues with the dataset. But the point is the data you just collected is raw data, which is often of poor quality. It's like crude oil. The first step is to collect the data, like extracting the crude oil. Crude oil is useless in that state. You cannot put crude oil in your car. You must refine and separate it using distillation to get products such as petrol or diesel. That's what you put in your car for use. Likewise, we need to do data-pre-processing to convert that raw data into its refined state. This takes time and can consume up to 80% of a data analyst's time to refine for analysis. But if you don't do it properly and give it all the time it needs, then it's likely that you will generate poor-quality findings. This is a crucial step because the analytic results are only as good as the data quality we throw in, as shown in Fig. 3.1.

Bonus
There are ways to mitigate this; poor data quality is often due to shoddy research practice or (if you're collecting via a web form or other online method) poor design on the form.

Implementing a machine learning model is fun and, easy and takes a couple of minutes. We will learn that in the next chapter. However, the most challenging part of being a data analyst is pre-processing the crude raw data into a more refined format. The steps depend on the type of data. For example, when working with text-based data, we need to do stemming and lemmatisation to generate the inflected words' root form. To understand stemming (used to reduce a word to its stem or root format) and lemmatisation (use it to reduce words to a normalised form), consider that we have the following words: "dance", "dancing", "danced" and "dancer". After applying stemming, all these words will be changed into "danc", and if you use lemmatisation instead, then all these words will be changed into "dance". It is important to perform stemming and lemmatisation while your data is text because the computer treats each of the following words "dance", "dancing", "danced" and "dancer" as unique words even though they are all talking about the practice of dance. So, if you are trying to learn about certain therapeutic practices or healing activities that helped patients with mental illnesses from their narratives (an example use case), then using stemming or lemmatisation can generate you a list. Comparatively, stemming is faster than lemmatisation because it chops words without knowing the context of the word in given sentences. In lemmatisation, it knows

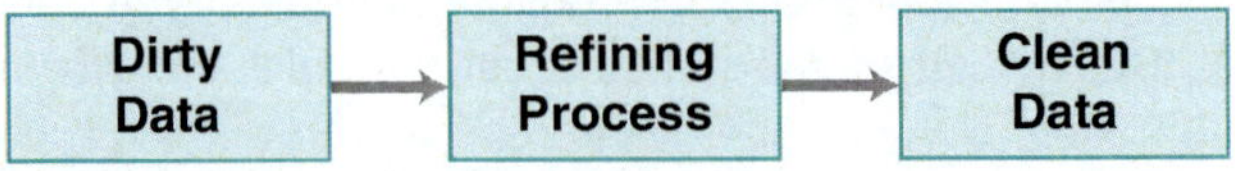

Fig. 3.1 Dirty data (raw) goes into and clean data (refined) comes out of the data-pre-processing step

the context before proceeding. Personally, I prefer lemmatisation. But there might be use cases where stemming makes more sense.

In summary, stemming and lemmatisation help the computer understand that the words "depress", "depression", "depressive" and "depressed" indicate the same word. Otherwise, it will treat each uniquely (which probably we don't want). But note that this is for text data. In this book, as mentioned before, we are focusing on numbers (probably the easiest to get started). So, let us take a peep into some of the most commonly used pre-processing steps applicable to numbers.

Broadly there are three types of data pre-processing (see Fig. 3.2. for an overview).

1. **Data cleaning** involves steps like taking care of the missing data and the outliers (or noisy data).
2. **Data transformation** involves converting the data from one format to another. It can convert categorical variables such as gender into binary variables where females are indicated by 1 and males by 0 or vice versa. There will be times when we would wish to rescale variable(s), such as converting temperature from Celsius to Fahrenheit, where we change the units of measurement of the data. This will also come under data transformation. Likewise, there are times when we compare measurements that have different units (e.g. age and salary). Variables measured at different scales do not contribute equally to the analysis and might create a bias. To prevent such biases from impacting our results, it is important to transform all the variables in the data on a single scale. We can use MinMax scaling (where we rescale the variables to a range of [0, 1]) or standardisation for such data transformation.

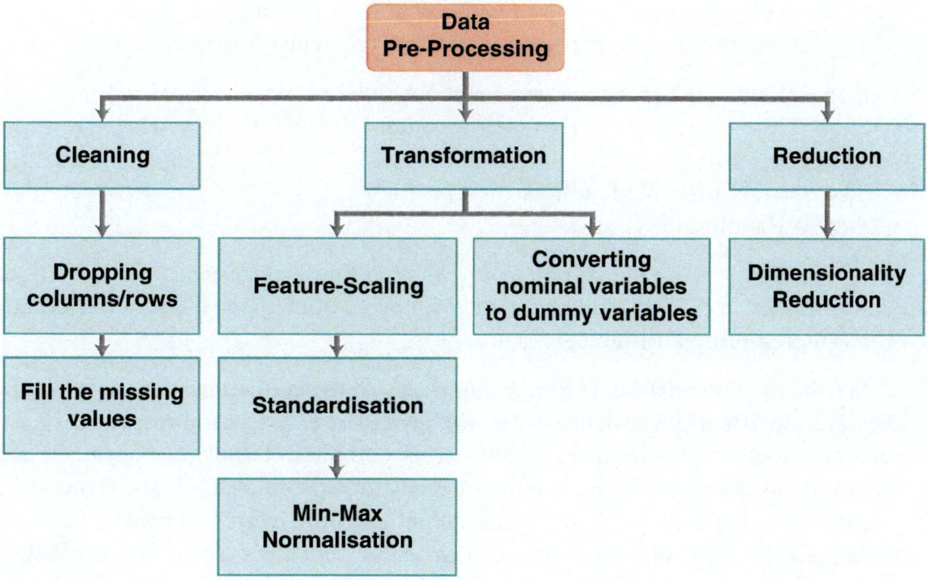

Fig. 3.2 Types of tasks involved within data-pre-processing

3. **Data reduction:** As mentioned above, not all independent variables contribute to your dependent variable. Subtracting such less relevant variables from the dataset increases storage efficiency, reduces costs and simplifies the model's interpretation.

In the following lessons, we will learn more about each of these and how to do them on your computer.

Bonus
As you will notice in the upcoming lessons, you can do some of these on your MS Excel or spreadsheet using REPLACE and REARRANGE (the columns) IN ASCENDING or DESCENDING order. That said, I strongly recommend you get used to using Python.

3.2 Data Cleaning

The raw dataset is similar to the crude oil extracted from the refineries. It's crude. It has impurities and therefore is unfit for use. Impurities in the dataset might consist of incorrect, corrupted, incorrectly formatted, duplicate or incomplete data. Data cleaning allows you to fix or remove them, and the result will be a refined dataset. Consequently, it results in a higher quality of data entering the machine learning model and, therefore, the highest-quality information in the predictive insights and better decision-making. To sum up, cleaning the data involves removing the data that does not belong in your dataset and preparing it for analysis.

The following steps are expected to provide you with a roadmap:

- Step 1: Remove duplicate or irrelevant observations
- Step 2: Fix structural errors (such as naming conventions, typos or incorrect capitalisation)
- Step 3: Filter unwanted outliers
- Step 4: Handling missing data:
  - Strategy 1: Drop the columns or rows with frequent missing values
  - Strategy 2: Fill the missing values with mean or median values of that column (representing a variable)

See the fictitious dataset below. A quick scan of the column headers suggests that the data consists of 23 patients. Thus, we have their age, sex and responses on two screening tools for depression and anxiety. If you did not know or notice, we used the two items from both the Patient Health Questionnaire-2 (PHQ-2) and Generalized Anxiety Disorder 2-item (GAD-2). The columns "Little interest or pleasure in doing things" and "Feeling down, depressed or hopeless" represent the items in the PHQ-2.

As you can see, it's inquiries about the frequency of depressed mood and anhedonia over the past 2 weeks. Likewise, the other two columns, "Feeling nervous, anxious or on edge" and "Not being able to stop or control worrying", come from the GAD-2. Together, they help us estimate depression and anxiety, respectively.

3.2.1 Problem 1: Duplicate Columns and Categorical Variables

The diagnosis of duplicate columns is simple. There are three steps (Fig. 3.3).

- Step 1: Import the relevant libraries, that is, pandas and numpy
- Step 2: Read the data file (which is in CSV format)
- Step 3: Use the `.head()` function to show the first few rows

The default is displaying the first five. But you can enter a numeric figure to specify how many rows you wish to display. For example, if instead of five, you wish to display seven rows of the data, then just enter seven within the `head()` and that many rows will be displayed (as we did in Fig. 3.4).

Result of Diagnosis The column AGE has been duplicated. The first row (PatientID 1) is duplicated as well. Sex is a categorical variable. Plus, we can see NaN values are indicating missing values.

If you noticed, the second line of code is

- `data = pd.read_csv("C:\\Users\\ghosh\\Downloads\\Tutorial_Data.csv", header=0, index_col=0)`

In it, there is `header=0`, and `index_col=0`. This tells Python to set the first row as column headers and the first column as index columns, respectively. The index column, the first column in the data file (which can be a spreadsheet), shall look like the participant ID or name (as can be found in most psychology datasheets).

Solution: Removing Duplicate Rows or Columns
Removing duplicate rows and columns must be removed once you have identified them. To be fair, while it is still possible to manually scan for the duplicate columns, it is pretty difficult to do the same with rows. The reason is simple. You might have 10–20 predictor variables to scan manually. But you will probably have 2000 rows

```
In [23]: import pandas as pd

In [24]: data = pd.read_csv("C:\\Users\\ghosh\\Downloads\\Tutorial_Data.csv", header=0, index_col=0)

In [27]: data.head() #For identifying duplicate rows/columns, missing values, and categorical variables
Out[27]:
```

PatientID	Sex	Age	Age.1	Little interest or pleasure in doing things	Feeling down, depressed or hopeless	Feeling nervous, anxious or on edge	Not being able to stop or control worrying
1	Male	36	36	3.0	3.0	3	3.0
1	Male	36	36	3.0	3.0	3	3.0
2	Male	200	200	NaN	NaN	1	2.0
3	Female	54	54	0.0	0.0	0	0.0
4	Male	38	38	1.0	1.0	1	2.0

Fig. 3.3 Codes to diagnose duplicate columns

```
In [28]: data.head(7) #For identifying duplicate rows/columns, missing values, and categorical variables
Out[28]:
```

PatientID	Sex	Age	Age.1	Little interest or pleasure in doing things	Feeling down, depressed or hopeless	Feeling nervous, anxious or on edge	Not being able to stop or control worrying
1	Male	36	36	3.0	3.0	3	3.0
1	Male	36	36	3.0	3.0	3	3.0
2	Male	200	200	NaN	NaN	1	2.0
3	Female	54	54	0.0	0.0	0	0.0
4	Male	38	38	1.0	1.0	1	2.0
5	Male	34	34	NaN	3.0	3	3.0
6	Female	37	37	1.0	1.0	2	NaN

Fig. 3.4 Line of code to display the first seven rows in the dataset

Duplicate Removal

```
In [29]: import pandas as pd

In [30]: data = pd.read_csv("C:\\Users\\ghosh\\Downloads\\Tutorial_Data.csv", header=0, index_col=0)

In [31]: data.head()
Out[31]:
```

PatientID	Sex	Age	Age.1	Little interest or pleasure in doing things	Feeling down, depressed or hopeless	Feeling nervous, anxious or on edge	Not being able to stop or control worrying
1	Male	36	36	3.0	3.0	3	3.0
1	Male	36	36	3.0	3.0	3	3.0
2	Male	200	200	NaN	NaN	1	2.0
3	Female	54	54	0.0	0.0	0	0.0
4	Male	38	38	1.0	1.0	1	2.0

```
In [32]: data = data.drop_duplicates(keep='first') #remove duplicate rows

In [33]: data = data.drop(columns =['Age.1']) #remove duplicate columns

In [34]: data.head()
Out[34]:
```

PatientID	Sex	Age	Little interest or pleasure in doing things	Feeling down, depressed or hopeless	Feeling nervous, anxious or on edge	Not being able to stop or control worrying
1	Male	36	3.0	3.0	3	3.0
2	Male	200	NaN	NaN	1	2.0
3	Female	54	0.0	0.0	0	0.0
4	Male	38	1.0	1.0	1	2.0
5	Male	34	NaN	3.0	3	3.0

Fig. 3.5 Codes to remove removing duplicate rows or columns

(representing respondents), which would make it time-consuming for you to manually identify the duplicates. The required codes are given in Fig. 3.5.

In the given example, a record (the first and second row) is flagged as duplicates because of an identical patient ID (i.e. PatientID 1), meaning it's the same person in two rows. That's the criterion being used. This is not mistaken with the thought that we're dropping anything where rows have identical values. Because it might be the case that two unique individuals have exactly the same responses to all items.

Step 1: Import pandas library.

Step 2: Read the data and assign it a variable "data" (most often, you will find people on the Internet using "`df`", but you can name it anything). The reason for using "df" is that in pandas, the common data structure is called a "data frame" – hence the common use of "df".

Step 3: Display the top five rows. This was optional because we already did this earlier in this lesson and identified the duplicate rows and columns. But I included it here (and at the end in step 6) anyway, just to demonstrate to you the data BEFORE and AFTER removing the duplicate rows and columns.

Step 4: Remove the duplicate rows using .drop_duplicates. I chose to keep the "first" one, but you can also choose to keep the "last one" – make sure you check if one record has the correct data, but the other is corrupt or erroneous. If both have correct data, it does not matter whether you keep the first or the last one. Otherwise, pick the record that has the correct data – otherwise, you'll produce poor results. If you noticed, I have not specified which row because even though you have identified that row 1 is duplicated, but that was possible because there were only 26 rows in this dataset. But imagine what if your dataset has 10,000 rows? Will you enter `data.head(10000)` and then manually scan for duplicate rows? Not so a feasible option. So, a more efficient approach is to tell Python to remove all the rows which have duplicates (retain the first and delete the rest).

Step 5: Remove the duplicate columns using `.drop(columns =['Age.1'])`. Now there can be 10,000 columns, and you might be thinking, what if we tell Python to remove all the duplicated columns? It is possible. But I recommend you go through your columns/variables manually and cherry-pick the most relevant ones with utmost care. I wouldn't do it manually if you have 10000 variables in your dataset. Still, instead, I will probably use correlation to find the ten variables with the highest correlation coefficient. Or, if that sounds ad hoc, then another idea would be to pick all the statistically significant variables to your variable of interest (target or dependent variable). Let's say you found 20 correlates of your target variable that are statistically significant. Now you can manually review and pick the variables which you think are of theoretical interest and use .drop_duplicates to remove the rest. You will learn more about it under the lesson on data reduction. But my point, for now, is do not blindly delete the duplicates because there might be irrelevant columns that had duplicates and remove just the duplicate. Not all the copies of that irrelevant column will probably do more harm than good.

As said before, if you have a large number of columns, then use something like correlation to narrow down your search. Once you have identified the most relevant or meaningful variables, retain them, and delete all others. So, out of 10,000, you might have got ten relevant variables. So, delete the remaining variables. If you ask me, I will tell you the lesser number of variables in your dataset, the better it is. This is important because of three reasons:

- **Reason 1:** The more variables or columns in the data, the sparser the dataset will get. This might introduce problems associated with the "curse of dimensionality". It has been observed that the average (expected) predictive power of a classifier or regressor first increases as the number of dimensions or features used is increased. But beyond a certain point, including additional parameters (that have been estimated) leads to a higher probability of error (Trunk, 1979). So, it is preferable to keep the number of columns as small as possible.

- **Reason 2:** The irrelevant variables might sometimes influence your findings (depending on your algorithm). For example, the potential impact of inter-relationships among the independent variables can confound the regression parameters and interpret them (Johnston et al., 2017). Therefore, only keep the variables you really need. Again, you will learn more about such inter-relations between independent variables in the lesson for data reduction.
- **Reason 3:** The more variables you include, the greater number of data observations (or higher the sample size) should be. In other words, the data size should increase proportionately as the number of variables increases. There is no rule written in stone, but it was proposed that there should be at least five training examples for each variable in the dataset (Theodoridis et al., 2008). So, if you have 30 respondents (i.e. rows or observations), then you should have no more than six variables (or columns). Do not worry if you don't understand a training example or the probabilities of error. As we advance into the next chapter on machine learning, we will come to them later. But for now, understand that keeping variables or columns that are irrelevant or ones that are not adding value is better to be removed along with the duplicate columns. You will find more discussion about this in the upcoming lesson on data reduction. There you will learn how to identify relevant variables and which ones can be deleted.

3.2.2 Problem 2: Outliers

At times you will find data points that differ significantly from other observations. For example, suppose you have gathered data about the weight of individuals in a classroom (assume same age and sex). In that case, you might notice that while almost everyone weighs around 62 kgs on average, one or two of them might be significantly overweight or underweight (e.g. 150 kgs or 24 kgs). These two peoples' weights can then be considered outliers. Such outliers can be a serious problem later on and might influence the results. Therefore, taking care of such outliers is required before analysing the data.

We can find or diagnose such outlier problems in the dataset using the `.describe()` on the data which employs descriptive statistics, such as central tendency, dispersion and shape of a dataset's distribution. Example codes can be found in Fig. 3.6.

In the above data table, the max age is 200 years, which is impossible and an outlier. However, the minimum and other maximum values look OK, that is, within the expected range.

There are two types of outliers: univariate and multivariate.

For Univariate

Diagnosis: One simple approach, as said above, is to use the describe function to see each variable's maximum and minimum values. Alternatively, you can use a box plot. For example, a box plot for a particular column can be created as shown

```
In [23]: import pandas as pd

In [24]: data = pd.read_csv("C:\\Users\\ghosh\\Downloads\\Tutorial_Data.csv", header=0, index_col=0)

In [25]: data.describe() #For identifying outliers
Out[25]:
```

	Age	Age.1	Little interest or pleasure in doing things	Feeling down, depressed or hopeless	Feeling nervous, anxious or on edge	Not being able to stop or control worrying
count	24.000000	24.000000	21.000000	23.0	24.000000	23.000000
mean	49.875000	49.875000	0.904762	1.0	1.333333	1.347826
std	35.079181	35.079181	1.044259	1.0	0.963087	0.982052
min	23.000000	23.000000	0.000000	0.0	0.000000	0.000000
25%	34.750000	34.750000	0.000000	0.0	1.000000	1.000000
50%	39.500000	39.500000	1.000000	1.0	1.000000	1.000000
75%	53.250000	53.250000	1.000000	1.0	2.000000	2.000000
max	200.000000	200.000000	3.000000	3.0	3.000000	3.000000

Fig. 3.6 Codes to diagnose outliers

```
In [54]: import pandas as pd

In [55]: data = pd.read_csv("C:\\Users\\ghosh\\Downloads\\Tutorial_Data.csv", header=0, index_col=0)

In [56]: import seaborn as sns
         sns.boxplot(x=data['Age'])
Out[56]: <AxesSubplot:xlabel='Age'>
```

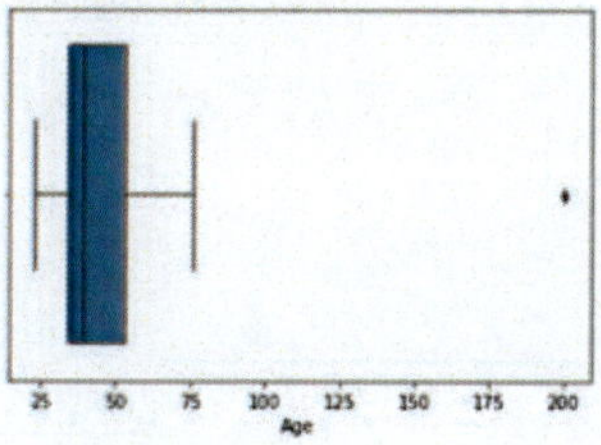

Fig. 3.7 Codes to create a box plot to diagnose outliers

in Fig. 3.7. We used Seaborn, a Python data visualisation library based on matplotlib. The reason I mentioned Seaborn is because

Solution: Set a range for your variable. For example, suppose you know that your study's inclusion criteria were 18–60 years. In that case, you can set this condition and delete all the respondents whose age has been noted as below 18 or above 60, as shown in Fig. 3.8.

For Multivariate

Diagnosis: You can use a scatter plot for a pair of columns to spot the outliers. Depending on your data and research aims, you can choose which two columns to choose for the scatter plot and select which values to consider outliers. The choice is on you. For example, see Fig. 3.9, I chose age and "feeling down…" as the two variables in this case.

As you can see, only two people responded to three, and their age were between 30 and 40 years. Likewise, two people between the age of 40 and 50 years responded two. These look like potential candidates for outliers.

```
In [63]: outlier = data[(data['Age'] >= 60)|(data['Age'] < 18)].index
         data.drop(outlier, inplace=True) #drops index rows from the data
         data['Age'].describe()

Out[63]: count    20.000000
         mean     39.250000
         std      10.636951
         min      23.000000
         25%      34.000000
         50%      37.500000
         75%      48.250000
         max      58.000000
         Name: Age, dtype: float64
```

Fig. 3.8 Codes to remove the outliers

```
In [60]: import matplotlib.pyplot as plt
         fig, ax = plt.subplots(figsize=(16,8))
         ax.scatter(data['Age'], data['Feeling down, depressed or hopeless'])
         ax.set_xlabel('Age')
         ax.set_ylabel('Feeling down, depressed or hopeless')
         plt.show()
```

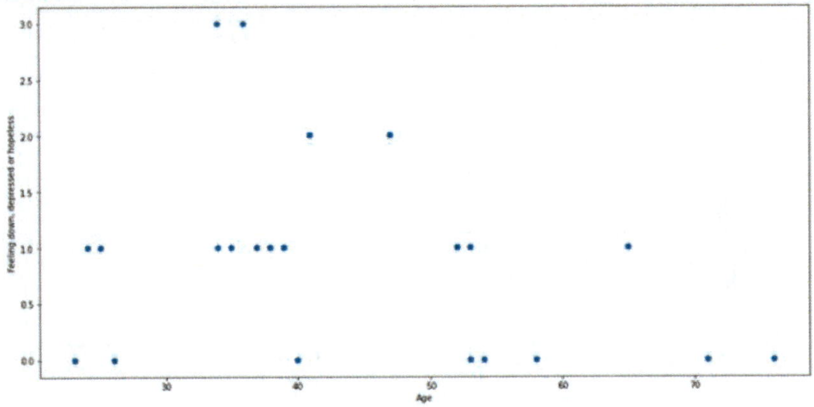

Fig. 3.9 Codes to create a scatter plot to diagnose outliers

Solution 1: You can delete the entire row

Solution 2: You can use the same codes mentioned above to delete the rows above a particular threshold. This is decided based on the topic and your domain knowledge.

3.2.3 *Problem 3: Missing Values*

Sometimes your survey respondent might have left a response blank, either intentionally or unintentionally. Such missing data points can cause trouble, so we need to identify and treat them.

Diagnosis: A little while back, I asked you to find the missing values using the manual scanning method (with `.head()`). But again, there can be 10,000 columns or 10,000 rows, making it inconvenient to scan it manually. So, we need a better way than that. The following line of code in Fig. 3.10 will help you achieve just that.

```
In [35]: print(data.isnull().sum())

         Sex                                            0
         Age                                            0
         Little interest or pleasure in doing things    3
         Feeling down, depressed or hopeless            1
         Feeling nervous, anxious or on edge            0
         Not being able to stop or control worrying     1
         dtype: int64
```

Fig. 3.10 Code to find out how many missing values are there in each column

```
In [41]: import pandas as pd

In [42]: data = pd.read_csv("C:\\Users\\ghosh\\Downloads\\Tutorial_Data.csv", header=0, index_col=0)

In [43]: print(data.isnull().sum())

         Sex                                            0
         Age                                            0
         Age.1                                          0
         Little interest or pleasure in doing things    3
         Feeling down, depressed or hopeless            1
         Feeling nervous, anxious or on edge            0
         Not being able to stop or control worrying     1
         dtype: int64

         Missing Values

In [44]: data['Little interest or pleasure in doing things'].fillna(data['Little interest or pleasure in doing things'].mean(), inplace=T

In [45]: data['Feeling down, depressed or hopeless'].fillna(data['Feeling down, depressed or hopeless'].mean(), inplace=True)

In [46]: data['Not being able to stop or control worrying'].fillna(data['Not being able to stop or control worrying'].mean(), inplace=Tru

In [47]: print(data.isnull().sum())

         Sex                                            0
         Age                                            0
         Age.1                                          0
         Little interest or pleasure in doing things    0
         Feeling down, depressed or hopeless            0
         Feeling nervous, anxious or on edge            0
         Not being able to stop or control worrying     0
         dtype: int64
```

Fig. 3.11 Codes to replace the missing values with the average ("mean") – alternative method to deleting the entire rows or columns

Once you know how many missing values are there in each of the variables in your dataset, it's easy to treat it.

Treatment: There are two solutions to deal with the problem of missing values.

Solution 1: Imputation

It means to replace the missing value with some value. In this case, we chose to replace the missing values with the mean of that column. So, after we had imported pandas and read the data file, we printed how many data points were missing in each column. After that, we imputed the missing values with .mean() for all three columns. See Fig. 3.11 for an example. Then we printed how many missing values are there in the columns like we did some time back. This time, we see there are zero missing values. Instead of mean, you can use median or something else. It's up to you.

Bonus

You must take care of something called "Data Leakage" while imputing or other data pre-processing methods. We will learn this under machine learning (page: 105, 112). Still, for now, it simply means that if you intend to split the data into two parts (which you will in machine learning), you will want those two subsets of the dataset to be independent and uninfluenced by one another. If you impute on the entire dataset, then the averaging for each variable will be made on the whole dataset. Then after that, if you divide the dataset into two parts, each part will have reminiscent of the whole dataset, and hence "data is said to be leaking". In simpler words, we would want both the parts of the dataset to be as pure as possible. So, first, we will divide the dataset into two parts and then perform the imputation on each part independently. That way, the average for each variable is estimated based on that part of the data only and not the entire dataset. It's ok not to understand this right now, and we will cover this more in depth later on in the machine learning chapter.

Solution 2: Drop or Delete the Entire Row

If you have a lot of data observations (e.g. rows/respondents) and you think you can afford to delete the entire row if there is a missing value in it, then you can simply use the following line of code to remove the entire rows (or people) from the dataset. See Fig. 3.12 for an example.

```
df.dropna(inplace = True)
```

To Sum Up (Table 3.1)

Now that we have identified potential issues or impurities within the data, it's time to clean them.

```
In [48]: import pandas as pd

In [49]: data = pd.read_csv("C:\\Users\\ghosh\\Downloads\\Tutorial_Data.csv", header=0, index_col=0)

In [50]: print(data.isnull().sum())

         Sex                                          0
         Age                                          0
         Age.1                                        0
         Little interest or pleasure in doing things  3
         Feeling down, depressed or hopeless          1
         Feeling nervous, anxious or on edge          0
         Not being able to stop or control worrying   1
         dtype: int64

In [52]: data.dropna(inplace = True) #Remove Missing Values

In [53]: print(data.isnull().sum())

         Sex                                          0
         Age                                          0
         Age.1                                        0
         Little interest or pleasure in doing things  0
         Feeling down, depressed or hopeless          0
         Feeling nervous, anxious or on edge          0
         Not being able to stop or control worrying   0
         dtype: int64
```

Fig. 3.12 Example code to drop or delete the entire row

Table 3.1 A summary of different data pre-processing approaches

Identifying the impurities in the data	Column/row	Solution
Duplicate column	Age	Delete/drop
Duplicate row	PatientID 1	Delete/drop
Missing value	PatientID 2, column = "Little interest or pleasure in doing things" and "feeling down, depressed or hopeless"	Delete/drop or impute/replace the blank space with a mean or median of the column (for example)
Categorical variable	Sex	Convert into dummy variable, such as FEMALE (where 0 represents NOT a female while 1 represents YES a female)
Outliers	Age = 200	Delete all the rows with outliers
Outliers	Age = 200	Delete all the rows with outliers

3.3 Data Transformation

There are different ways to transform the dataset to make it more meaningful for analysis.

3.3.1 Converting Categorical Variables into Numeric Variables

Python can process text data, but you need to transform it into numeric form. One approach is to convert the categorical variables such as sex: male/female into a dummy variable (Dummy_Male) where 1 represents male and 0 represents female. Figure 3.13 demonstrates the codes for performing this.

Bonus
The code cell in Fig. 3.14a is … same as typing into two different cells in Jupyter as is shown in Fig. 3.14b.

3.3.2 Converting Continuous Variables into Categorical Variables

Suppose you have a variable that indicates the pain intensity (on a scale of 0–100). Sometimes, you might want to convert this continuous data (0–100) into a categorical variable (e.g. no pain, mild pain, moderate pain and extreme pain). Accordingly, you assigned people who rated their pain between 0 and 25 as "no pain". Likewise,

```
In [6]: import pandas as pd

In [7]: data = pd.read_csv("C:\\Users\\ghosh\\Downloads\\Tutorial_Data.csv", header=0, index_col=0)
        data.head()

Out[7]:
```

PatientID	Sex	Age	Age.1	Little interest or pleasure in doing things	Feeling down, depressed or hopeless	Feeling nervous, anxious or on edge	Not being able to stop or control worrying
1	Male	36	36	3.0	3.0	3	3.0
1	Male	36	36	3.0	3.0	3	3.0
2	Male	200	200	NaN	NaN	1	2.0
3	Female	54	54	0.0	0.0	0	0.0
4	Male	38	38	1.0	1.0	1	2.0

```
In [8]: data = pd.get_dummies(data, columns=['Sex'], drop_first=True, prefix='Dummy')
        data.head()

Out[8]:
```

PatientID	Age	Age.1	Little interest or pleasure in doing things	Feeling down, depressed or hopeless	Feeling nervous, anxious or on edge	Not being able to stop or control worrying	Dummy_Male
1	36	36	3.0	3.0	3	3.0	1
1	36	36	3.0	3.0	3	3.0	1
2	200	200	NaN	NaN	1	2.0	1
3	54	54	0.0	0.0	0	0.0	0
4	38	38	1.0	1.0	1	2.0	1

Fig. 3.13 Code to convert categorical variable into a dummy variable

```
In [7]: data = pd.read_csv("C:\\Users\\ghosh\\Downloads\\Tutorial_Data.csv", header=0, index_col=0)
        data.head()
```

Fig. 3.14a Python Codes to read the CSV file and then display the first five rows of dataset (using one Jupyter cell)

```
In [ ]: data = pd.read_csv("C:\\Users\\ghosh\\Downloads\\Tutorial_Data.csv", header=0, index_col=0)

In [ ]: data.head()
```

Fig. 3.14b Python codes to read the CSV file and then display the first five rows of dataset (using two Jupyter cells)

25–50 represents mild pain, 50–75 represents moderate pain, and 75–100 represents extreme pain. This is called **discretisation.**

Two more familiar examples of **discretisation** would probably be the conversion of:

- Age from continuous data (18–60 years) into young adults (18–30 years), middle age (31–50 years) and elderly (51–60 years)
- Salary from continuous data (£25k PLUS_SPI) into low-income, middle-income and high-income groups

If you want to divide the data into equal parts, then you can use the code below. In this case, I have divided age into three equal groups, and you might wish to divide it into two or four. All you need to do is change the number of bins to get your desired outcome. See Fig. 3.15, for example.

If you want to convert the continuous variable, such as age, into a categorical variable, then try the following code (Fig. 3.16) and spot how it differs from the previous code. See the output result for the difference.

```
In [33]: categorising_age = pd.cut (x = data.Age, bins = 3)
         categorising_age

Out[33]: PatientID
         1     (34.667, 46.333]
         3       (46.333, 58.0]
         4     (34.667, 46.333]
         5     (22.965, 34.667]
         6     (34.667, 46.333]
         7       (46.333, 58.0]
         8       (46.333, 58.0]
         9     (22.965, 34.667]
         10    (34.667, 46.333]
         11    (22.965, 34.667]
         12      (46.333, 58.0]
         13    (34.667, 46.333]
         14    (34.667, 46.333]
         15    (22.965, 34.667]
         16      (46.333, 58.0]
         17    (22.965, 34.667]
         18      (46.333, 58.0]
         19    (22.965, 34.667]
         20    (34.667, 46.333]
         Name: Age, dtype: category
         Categories (3, interval[float64]): [(22.965, 34.667] < (34.667, 46.333] < (46.333, 58.0]]

In [34]: data.head()

Out[34]:
```

PatientID	Age	Little interest or pleasure in doing things	Feeling down, depressed or hopeless	Feeling nervous, anxious or on edge	Not being able to stop or control worrying	Dummy_Male
1	36	3.000000	3.0	3	3.000000	1
3	54	0.000000	0.0	0	0.000000	0
4	38	1.000000	1.0	1	2.000000	1
5	34	0.882353	3.0	3	3.000000	1
6	37	1.000000	1.0	2	1.333333	0

Fig. 3.15 Codes to convert continuous variables into categorical variables (method 1)

```
In [97]: my_labels = ['Young', 'Middle-aged', 'Elderly']

In [105]: data['Age_Categories'] = pd.cut (x = data.Age, bins = 3, labels = my_labels)
          #categorising_age

In [106]: data.head()

Out[106]:
```

PatientID	Age	Little interest or pleasure in doing things	Feeling down, depressed or hopeless	Feeling nervous, anxious or on edge	Not being able to stop or control worrying	Dummy_Male	Age_Categories
1	36	3.000000	3.0	3	3.000000	1	Middle-aged
3	54	0.000000	0.0	0	0.000000	0	Elderly
4	38	1.000000	1.0	1	2.000000	1	Middle-aged
5	34	0.904762	3.0	3	3.000000	1	Young
6	37	1.000000	1.0	2	1.347826	0	Middle-aged

Fig. 3.16 Codes to convert continuous variables into categorical variables (method 2)

Note: `pd.cut (x = data.Age, bins = 3, labels = my_labels)`

The "`data['Age_Categories'] =`" adds a new column with age categories in the dataset. Likewise, the .cut on pandas allows Python to group the continuous data into different categories. The "`x =`" specifies which column or variable you wish to categorise, and "`labels =`" asks how it should label the different categories. For example, if you have set `bins = 10` (which suggests you want ten divisions for that variable), you should provide ten labels `my_labels= ['label_1', 'label_2',….'label_10']`. Of course, the variable "`my_label`" can be named anything really (see the lesson on variables in the previous chapter of this book).

```
In [108]: data = pd.get_dummies(data, columns=['Age_Categories'], drop_first=True, prefix='Dummy')
          data.head()
Out[108]:
```

PatientID	Age	Little interest or pleasure in doing things	Feeling down, depressed or hopeless	Feeling nervous, anxious or on edge	Not being able to stop or control worrying	Dummy_Male	Dummy_Middle-aged	Dummy_Elderly
1	36	3.000000	3.0	3	3.000000	1	1	0
3	54	0.000000	0.0	0	0.000000	0	0	1
4	38	1.000000	1.0	1	2.000000	1	1	0
5	34	0.904762	3.0	3	3.000000	1	0	0
6	37	1.000000	1.0	2	1.347826	0	1	0

Fig. 3.17 Codes to convert a categorical variable into a dummy variable

```
In [149]: my_labels = ['Young', 'Middle-aged', 'Elderly']
          data['Age_Categories'] = pd.qcut (x = data.Age, q = 3, labels = my_labels)
          #categorising_age

In [150]: data.head()
Out[150]:
```

PatientID	Age	Little interest or pleasure in doing things	Feeling down, depressed or hopeless	Feeling nervous, anxious or on edge	Not being able to stop or control worrying	Dummy_Male	Dummy_Middle-aged	Dummy_Elderly	Age_Categories
1	36	3.000000	3.0	3	3.000000	1	1	0	Middle-aged
3	54	0.000000	0.0	0	0.000000	0	0	1	Elderly
4	38	1.000000	1.0	1	2.000000	1	1	0	Middle-aged
5	34	0.904762	3.0	3	3.000000	1	0	0	Young
6	37	1.000000	1.0	2	1.347826	0	1	0	Middle-aged

Fig. 3.18 Codes to convert a continuous variable into a categorical variable using **quantiles**

Suppose you try to run the above dataset without converting the categorical age variable into a dummy variable. In that case, Python will throw up an error. Let's run the `.get_dummies` once again and see how it changes the dataset. See Fig. 3.17 for the codes.

Did you see? This time three dummy variables have been created.

Now, what if you want to create the categories based on rank or based on sample quantiles?

It's really simple. Instead of `pd.cut` use `pd.qcut`. `.qcut` is the quantile-based discretisation function. One difference is that instead of "`bins =`", we set "`q=`" here. When q = 3, as in this case, it is called terciles: q= [0, 1/3, 2/3, 1]. It indicates either of the two points that divide an ordered distribution into three parts, each containing a third of the population. You might be more familiar with the concept of quartile (q = 4), which means any of the three points that divide an ordered distribution into four parts, each containing a continuous variable in terms of that many groups of percentiles. See Fig. 3.18 for the codes.

3.3.3 *Feature Scaling*

[A soft reminder: "Feature" for machine learning practitioners is what "independent variable" is for statisticians and social scientists].

Table 3.2 Sample dataset

Participant name	Age (years)	Salary (£)
Robin	19	21,500
Martha	28	53,000
Louise	36	25,000
Tracy	58	40,000

Table 3.3 Magnitude and unit of two variables

Variables	Average magnitude	Unit
Age	30	Years
Salary	400,000	£

Imagine a particular dataset (Table 3.2). It has two columns: AGE and SALARY for 30 research participants.

You need to know that every variable has two components: magnitude and a unit (Table 3.3).

In this case, the unit of age is in years, and that of salary is in currency (e.g. £ and $). In terms of magnitude, age ranges between 18 and 60 years and salary ranges between £20,000 and £50,000.

We, humans, understand that the units of age and salary are different. But machine learning algorithms are not good at understanding such units of values. It tends to weigh bigger values (e.g. salary) and higher and considers the relatively small values (e.g. age) as the lower values. In this example, the machine learning algorithm will treat salary greater than age leading to inappropriate predictions.

We consider bringing all values (data) to the same magnitudes to prevent such mishaps. This is called feature scale (because we are rescaling the features) or data transformation (because we are transforming the data values).

To put it in simple words: we do feature scaling because many machine learning algorithms perform better when features (variables) are on a relatively similar scale or are close to normally distributed.

Where Is Feature Scaling Needed?

Not all machine learning algorithms require feature scaling. See Table 3.4 for example.

P.S. It is ok not to be familiar with many of these algorithms at this stage, and we will learn more about them in the upcoming chapter. You can refer to this table whenever you learn a new algorithm. Just to check if it needs feature scaling.

How to Do Feature Scaling?

There are two popular strategies: MinMaxScaler and StandardScaler (standardisation). Let me demonstrate these to you right away. See Fig. 3.19 for the codes.

Table 3.4 Shortcut guide to whether your algorithm requires feature scaling

Feature scaling	Machine learning algorithms
Needed	Linear regression, logistic regression, support vector machine (SVM), neural networks, k-nearest neighbours algorithm (k-NN), k-means and other distance-based algorithms or gradient descent-based algorithms
Not needed	Decision tree, random tree and other tree-based algorithms

```
from sklearn import preprocessing
import pandas as pd
dataset = pd.read_csv("C:\\Users\\ghosh\\Downloads\\K-MeansToy.csv", index_col = 'Name', encoding='latin1')
#Standardisation
scaler = preprocessing.StandardScaler()
names = dataset.columns
fitting_to_dataset = scaler.fit_transform(dataset)
scaled_data = pd.DataFrame(fitting_to_dataset, columns=names)
print (scaled_data.head(10))
scaled_data.to_excel('Standardised_Data.xlsx', engine='xlsxwriter')
```

Fig. 3.19 Codes to perform feature scaling

Normalisation

```
In [109]: #Import
          from sklearn.preprocessing import MinMaxScaler

          #Normalise the dataset
          scaler = MinMaxScaler()
          normalised_data = pd.DataFrame(scaler.fit_transform(data))

          #Print the normalised data
          print (normalised_data.head())

          #Exporting the normalised data to your computer
          normalised_data.to_excel('Normalised_Dataset_Tutorial.xlsx', engine='xlsxwriter')

          0         1         2         3         4    5    6    7
0  0.371429  1.000000  1.000000  1.000000  1.000000  1.0  1.0  0.0
1  0.885714  0.000000  0.000000  0.000000  0.000000  0.0  0.0  1.0
2  0.428571  0.333333  0.333333  0.333333  0.666667  1.0  1.0  0.0
3  0.314286  0.301587  1.000000  1.000000  1.000000  1.0  0.0  0.0
4  0.400000  0.333333  0.333333  0.666667  0.449275  0.0  1.0  0.0
```

Fig. 3.20 Codes to perform MinMaxScaler

MinMax Scaling

It refers to the process of rescaling a variable value with a distribution (fixed) value between 0 and 1. The benefit of having this bounded range is that there will be smaller standard deviations, limiting the effect of outliers. Otherwise, outliers can affect several machine learning algorithms' performances, such as logistic regression and k-nearest neighbour (KNN) algorithm.

Refer to Fig. 3.20. Did you notice how MinMaxScaler changed the original dataset?

Bonus: While browsing through the codes on the Internet, you will regularly come across some codes using the Normalizer function. You might see it as follows:

```
from sklearn.preprocessing import Normalizer
transformer = Normalizer().fit(X)
```

The `Normalizer()` is not the same as `MinMaxScaler ()`. We used `MinMaxScaler ()` which rescales the dataset between 0 and 1. On the other hand, the Normalizer rescales the vector for each sample to have a unit norm, independently of the distribution of the samples.

Standardisation (or Z-Score Normalisation) To rescale, a variable value must have a distribution with mean value = 0 and variance equals 1. If you are a behavioural scientist, chances are you have a background in statistics and should be familiar with z-score. It is a prerequisite for some algorithms such as artificial neural network, naive Bayes and k-nearest neighbour (KNN) algorithm (Fig. 3.21).

In the previous lesson, you learnt about MinMaxScaler and, in this lesson, about Z-score normalisation. Now it's either-or, and it makes no sense to do both together. Then the next obvious question that might pop into your head is:

Which one to opt for?

The answer is not straightforward. As with any machine learning parameter selection, you have to try out both the options (at different times) to see which is improving your algorithm's performance.

Did you notice that the headers have changed to numeric?

That's something that happens, and you have to manually copy-paste the headers from the original dataset to the scaled or standardised dataset using Excel or spreadsheet. See Fig. 3.22 (state A) and how it has been changed in Fig. 3.23 (state 2).

Note that we generated the output file in Excel format. I recommend converting it into CSV format before using the data file for further analysis. I am using MS Excel, and it's really simple to convert from Excel to CSV format. Go to Export>Change File Type>CSV>Save As (refer to Fig. 3.24).

There are times when you might get an error message. In such cases, you need to re-look the codes. For example, one such error can be seen in Fig. 3.25. The error

Standardisation

```
In [146]: #Import
          from sklearn.preprocessing import StandardScaler

          #Standardising the data
          scaler = StandardScaler()
          standardised_data = pd.DataFrame(scaler.fit_transform(data))

          #Print the standardised data
          print (standardised_data.head(10))

          #Exporting the standardised dataset in your computer
          standardised_data.to_excel('Standardised_Dataset_Tutorial.xlsx', engine='xlsxwriter')
```

```
          0         1         2         3         4         5         6  \
0 -0.322455  2.322154  2.179449  1.725324  1.815365  1.471960  1.309307
1  1.374153 -0.971233 -1.089725 -1.347910 -1.453788 -0.679366 -0.763763
2 -0.133943  0.126563  0.000000 -0.323498  0.725647  1.471960  1.309307
3 -0.510967  0.022011  2.179449  1.725324  1.815365  1.471960 -0.763763
4 -0.228199  0.126563  0.000000  0.700913  0.014962 -0.679366  1.309307
5  0.714361  1.224359  1.089725  0.700913  1.815365 -0.679366 -0.763763
6  1.185641  0.022011  0.000000  0.700913 -0.364071 -0.679366 -0.763763
7 -1.453526  0.126563  0.000000 -0.323498  0.725647 -0.679366 -0.763763
8 -0.416711  0.126563  0.000000 -0.323498 -0.364071 -0.679366  1.309307
9 -0.510967  0.126563  0.000000 -0.323498 -0.364071 -0.679366 -0.763763

          7
0 -0.679366
1  1.471960
2 -0.679366
3 -0.679366
4 -0.679366
5  1.471960
6  1.471960
7 -0.679366
8 -0.679366
9 -0.679366
```

Fig. 3.21 Codes to standardise the dataset

	A	B	C	D	E	F	G	H	I
1		0	1	2	3	4	5	6	7
2	0	0.371429	1	1	1	1	1	1	0
3	1	0.885714	0	0	0	0	0	0	1
4	2	0.428571	0.333333	0.333333	0.333333	0.666667	1	1	0
5	3	0.314286	0.301587	1	1	1	1	0	0
6	4	0.4	0.333333	0.333333	0.666667	0.449275	0	1	0
7	5	0.685714	0.666667	0.666667	0.666667	1	0	0	1
8	6	0.828571	0.301587	0.333333	0.666667	0.333333	0	0	1
9	7	0.028571	0.333333	0.333333	0.333333	0.666667	0	0	0
10	8	0.342857	0.333333	0.333333	0.333333	0.333333	0	0	0
11	9	0.314286	0.333333	0.333333	0.333333	0.333333	0	0	0
12	10	0.857143	0	0.333333	0	0.333333	0	0	1
13	11	0.514286	1	0.666667	0.666667	0.666667	0	1	0
14	12	0.485714	0.333333	0	0.333333	0.333333	0	1	0
15	13	0	0	0	0.333333	0.333333	0	0	0

Fig. 3.22 State A (MinMax scaled data, notice Python removes all the headers)

	A	B	C	D	E	F	G	H	I	J
1		Age	Little interest	Feeling down, dep	Feeling nervous,	Not being able to	Dummy_Male	Dummy_Middle-aged	Dummy_Elderly	
2	0	0.371429	1	1	1	1	1	1	0	
3	1	0.885714	0	0	0	0	0	0	1	
4	2	0.428571	0.333333333	0.333333333	0.333333333	0.666666667	1	1	0	
5	3	0.314286	0.301587302	1	1	1	1	0	0	
6	4	0.4	0.333333333	0.333333333	0.666666667	0.449275362	0	1	0	
7	5	0.685714	0.666666667	0.666666667	0.666666667	1	0	0	1	
8	6	0.828571	0.301587302	0.333333333	0.666666667	0.333333333	0	0	1	
9	7	0.028571	0.333333333	0.333333333	0.333333333	0.666666667	0	0	0	
10	8	0.342857	0.333333333	0.333333333	0.333333333	0.333333333	0	0	0	
11	9	0.314286	0.333333333	0.333333333	0.333333333	0.333333333	0	0	0	
12	10	0.857143	0	0.333333333	0	0.333333333	0	0	1	
13	11	0.514286	1	0.666666667	0.666666667	0.666666667	0	1	0	
14	12	0.485714	0.333333333	0	0.333333333	0.333333333	0	1	0	
15	13	0	0	0	0.333333333	0.333333333	0	0	0	

Fig. 3.23 State B (MinMax scaled data – insert the headers manually, or just copy-paste the headers in the Excel or spreadsheet file from the original dataset)

came up because I wrote it using the `read_csv()` method, but pandas also has a `read_excel()` method that works great. So be mindful of such small mistakes.

If you used the same code (`read_csv()`) but had a CSV file instead of an Excel file, then it would have run smoothly without any error message (Fig. 3.26).

3.4 Data Reduction

We all know that the more decent quality data we can collect, the more information we have, and the better it is to make inferences or predictions. Often what we mean is more the number of observations (or respondents), the better it is. But we don't

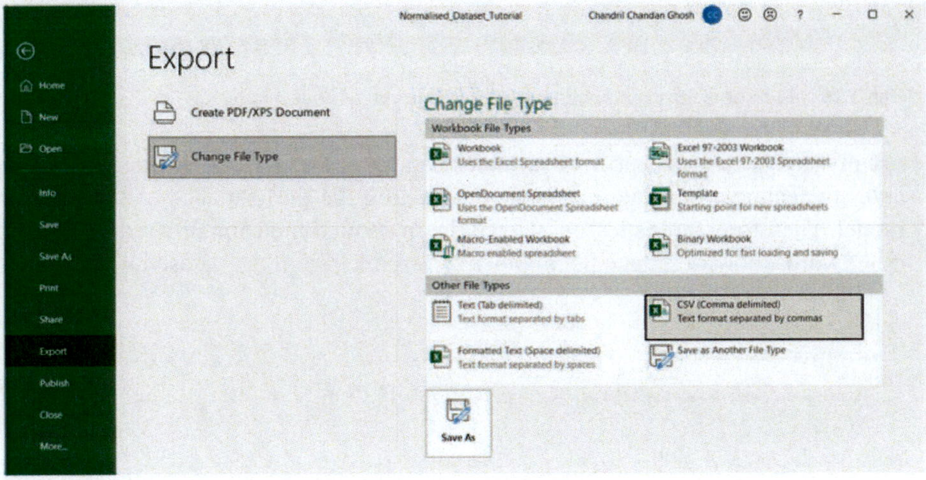

Fig. 3.24 Option on MS Excel to convert the file into CSV format

Fig. 3.25 An example of an error message thrown up by Python

```
In [255]: import pandas as pd
          data = pd.read_csv("C:\\Users\\ghosh\\Downloads\\Normalised_Dataset_Tutorial.csv", header=0, index_col=0)
```

Fig. 3.26 No error is thrown up while executing this cell of code

usually say more the number of variables better. As we increase the number of variables or features, our data gets sparse, affecting the performance of the analytics model. Therefore, in this lesson, we will learn about dimensionality reduction, or in other words, how to reduce the number of variables from the dataset.

Bonus
A lot of bad data can produce worse results than a smaller amount of quality data.

A typical dataset has two aspects: the number of participants (n) and variables or predictors (p). There is no hard and fast rule to this. Still, often we will see that the researcher has represented the rows as people and the columns as variables or features. Or that's how I personally structure my datasets. Accordingly, if there are 30 rows, then there are 30 research participants. Generally, the more rows we have, the more the number of people we collected data from, and the merrier it is. We learnt this in the basics of statistics. But the more the number of columns we have to analyse, the more the number of variables, the more storage space it will consume, and the longer the computation time. Space and time issues will become more prominent when we deal with big data.

At the same time, there might be many redundant variables that are not contributing to our prediction. Instead, those columns make the data sparse, leading to all sorts of problems called the "curse of dimensionality". Search for this dramatic phrase on the Internet, you will get more in-depth information about this problem. But for now, know that the dimensionality reduction techniques allow us to enhance the processing speed and reduce the time and effort required to extract valuable information (Ayesha et al., 2020). So, to sum up, delete the columns or variables you do not need and those not contributing to the prediction task.

Bonus 1
Variables, features and dimensions are used interchangeably in the field, but all mean the exact same thing.

Bonus 2
Note that all the steps in pre-processing are not compulsory. It solely depends on your dataset and what you want to achieve. Just because I am teaching MinMaxScaler (in the previous lesson) and dimensionality reduction strategies (in the current lesson) does not mean you have to do them all all the time. For example, as mentioned in the previous lesson, tree-based machine learning algorithms do not require feature rescaling. Likewise, suppose you have sufficient reason to believe that all the columns or variables are important for you. In that case, there is no point in reducing/deleting them.

How Do You Select Which Variables or Features to Delete?
First of all, we are talking about the independent variables (predictors) here. We will try to predict the predictors' dependent or target variable (predicted). So I would say the lesser the number of predictors, the cleaner and leaner your model will be. If you are unsure whether to delete a particular predictor from the data and if literature or common sense is not helping, you can opt for dimensionality reduction strategies and get the answer to your question.

In my experience, many of the data in psychology do not require dimension reduction. But many researchers opt for that. So let me give you a familiar example and, in doing so, demonstrate to you the first strategy of dimension reduction.

3.4.1 Strategy 1

Summing Up Multiple Variables to Create a New Variable Suppose a researcher collects data on a depression questionnaire called PHQ-9. There are nine symptoms of depression, and the subjects' responses to each of those symptoms (item) are entered on the spreadsheet. See Fig. 3.27 for the dataset. So, the researcher has nine columns in the dataset. Now there are times when the researcher's interest might lie in just the depression sum score, not the individual symptoms. In such cases, they will sum up all the item responses for all the subjects and create a new column called "depression severity score". Then she will probably delete all the columns that included item-level responses. You see, earlier we had nine columns, and now we have only one column. Psychologists will be more familiar with this dimension reduction strategy than many machine learning practitioners with a computer science background.

This approach is similar to **concept hierarchy generation**. You might argue that they are the same. Here is how it goes: Suppose we have got data about the person's city of origin. We have got people from Belfast, Dombivli, Kolkata, Delhi and London. They might also have entered their postcode or street address. So, there are three variables: city, street and postcode. Sometimes we want to reduce the number of variations by grouping similar variables together. Here, we could do so by converting the "city, street and postcode" to just one variable "country", which enables

Method 1: Summing up columns

```
In [204]: data['Depression'] =  data[['Little interest or pleasure in doing things', 'Feeling down, depressed or hopeless']].sum(axis=1)
          data.head(10)
```

Out[204]:

PatientID	Age	Little interest or pleasure in doing things	Feeling down, depressed or hopeless	Feeling nervous, anxious or on edge	Not being able to stop or control worrying	Dummy_Male	Dummy_Middle-aged	Dummy_Elderly	Depression
1	36	3.000000	3.0	3	3.000000	1	1	0	6.000000
3	54	0.000000	0.0	0	0.000000	0	0	1	0.000000
4	38	1.000000	1.0	1	2.000000	1	1	0	2.000000
5	34	0.904762	3.0	3	3.000000	1	0	0	3.904762
6	37	1.000000	1.0	2	1.347826	0	1	0	2.000000
7	47	2.000000	2.0	2	3.000000	0	0	1	4.000000
8	52	0.904762	1.0	2	1.000000	0	0	1	1.904762
9	24	1.000000	1.0	1	2.000000	0	0	0	2.000000
10	35	1.000000	1.0	1	1.000000	0	0	0	2.000000
11	34	1.000000	1.0	1	1.000000	0	0	0	2.000000

Fig. 3.27 Codes showing how to sum up multiple variables to create a new variable

Method 2: Delete the columns with missing values

```
In [ ]: data = data.dropna(axis='columns', how='any', thresh=len(df)*0.8) #Drop columns with more than 20% missing values
```

Fig. 3.28 Codes to remove the variables with "too many" missing values

us to mark our respondents in terms of India (Dombivli, Kolkata, Delhi) or the United Kingdom (London and Belfast).

3.4.2 Strategy 2

Removing the Variables with "Too many" missing values If there are columns with too many **missing values**, then delete those columns. How many is too many depends on your personal judgement and research objective? You can set a threshold value, and if any variable's percentage of missing values is more than that threshold, you can drop the variable. Refer to the codes presented in Fig. 3.28.

3.4.3 Strategy 3

Remove the Variables with Too Little Variance Check for the **variance** for each column. Columns with low variance can probably be deleted from the dataset. Again, as with missing values, you can set a threshold and drop the variables with a lower variance. But note that you normalise or rescale the data before estimating the variances. The reason for deleting such variables with low variance is that such variables are not distinguishing between your respondents well enough. Imagine, on a Likert scale of 0 to 5 of a variable, all your respondents reported 0. Then from a data analysis point of view, it is not adding any useful information to predict your target variable. See Fig. 3.29 for the sample codes to achieve that.

```
In [52]: import pandas as pd
         data = pd.read_csv("C:\\Users\\ghosh\\Downloads\\Normalised_Dataset_Tutorial.csv", header=0, index_col=0)
```

Method 3: Delete columns with low variance.

```
In [53]: from sklearn.feature_selection import VarianceThreshold
         variance_filter = VarianceThreshold(threshold=0)
         variance_filter.fit(data)
Out[53]: VarianceThreshold(threshold=0)
```

Fig. 3.29 Codes to remove the variables with too little variance

3.4.4 Strategy 4

Filter Using Correlation If you have a statistics background, you must be familiar with the term "multicollinearity". This refers to a "problem" that affects the coefficients and p-values. Multicollinearity occurs when two or more independent or predictor variables are highly correlated with each other in a regression model. This is problematic because regression analysis aims to isolate the relationship between each independent variable and the dependent variable. However, when the independent variables are correlated with each other, it suggests that changes in one predictor are associated with changes in another predictor. It becomes difficult for the model to independently estimate the relationship between each predictor (independent variable) and the dependent variable because some independent variables (predictors) tend to change together. In other words, it makes it impossible to distinguish the individual effects of the independent or predictor variables on the dependent variable. So, we perform a Pearson correlation as a preliminary analysis for identifying "multicollinearity" in regression. See Fig. 3.30 for the codes and heat map plot. We would ignore the ones with small or moderate correlations. Still, we will remove one variable from the pair of variables with high correlation coefficients – an exception to the fact that when multicollinearity is a problem lies with the goal of your analysis.

For example, suppose you are a machine learning practitioner developing the model for optimising an app functionality. In that case, your primary goal is to make predictions. You are not interested to understand the role of each independent variable or predictor. In that case, you don't need to reduce severe multicollinearity. But as a psychologist, most often, you will be interested in the latter, that is, understanding the contribution or role of each predictor. Therefore, you should check and remove highly correlated predictors. Suppose you did not notice the above discussion. In that case, we are discussing correlations within predictor variables (not with the target or dependent variable).

Take a note to normalise the data before estimating the correlations.

Bonus

If you want to export and save the image file and the correlation matrix in a CSV format, add a new cell after running the previous cell (shown in Fig. 3.30), as presented in Fig. 3.31.

Method 4: Deleting highly correlated columns

```
In [62]: import matplotlib.pyplot as plt

         corr_Matrix = data.corr()

         ax = sns.heatmap(corr_Matrix, annot=True)
         bottom, top = ax.get_ylim()
         ax.set_ylim(bottom + 0.4, top - 0.4)

         plt.show()
```

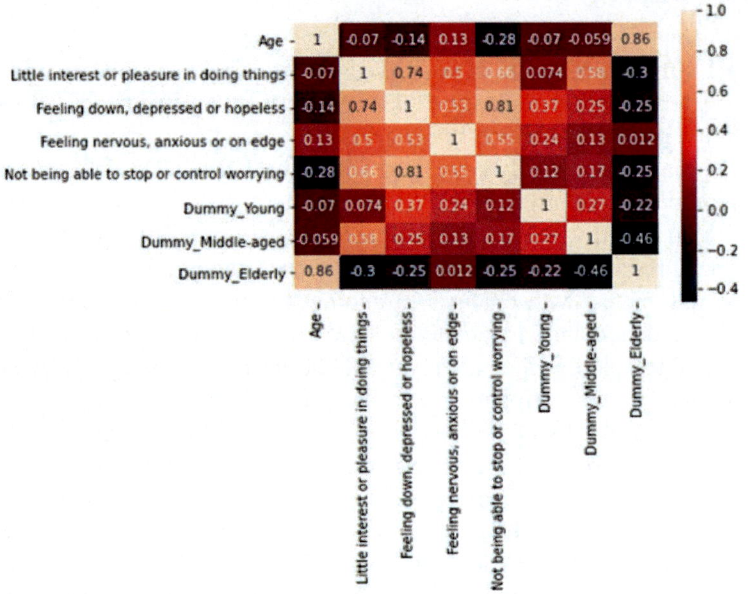


```
<Figure size 432x288 with 0 Axes>
```

Fig. 3.30 Code to display correlation coefficients between pairs of variables using a heat map

```
In [58]: plt.savefig('correlation_matrix_image.png')

         data.corr().to_csv(r"C:\Users\ghosh\Downloads\correlation matrix.csv")

         <Figure size 432x288 with 0 Axes>
```

Fig. 3.31 Code to export the correlation matrix in a CSV file

Practise Exercise The code in Figs. 3.30 and 3.32 has a missing line of code. So it probably won't work as is. Take a moment to think, what is that?

- **Hint:** Run this code to see what the error message is telling you.

Method 4: Deleting highly correlated columns

```
In [62]: import matplotlib.pyplot as plt

         corr_Matrix = data.corr()

         ax = sns.heatmap(corr_Matrix, annot=True)
         bottom, top = ax.get_ylim()
         ax.set_ylim(bottom + 0.4, top - 0.4)

         plt.show()

         plt.savefig('correlation_matrix_image.png')

         data.corr().to_csv(r"C:\Users\ghosh\Downloads\correlation matrix.csv")
```

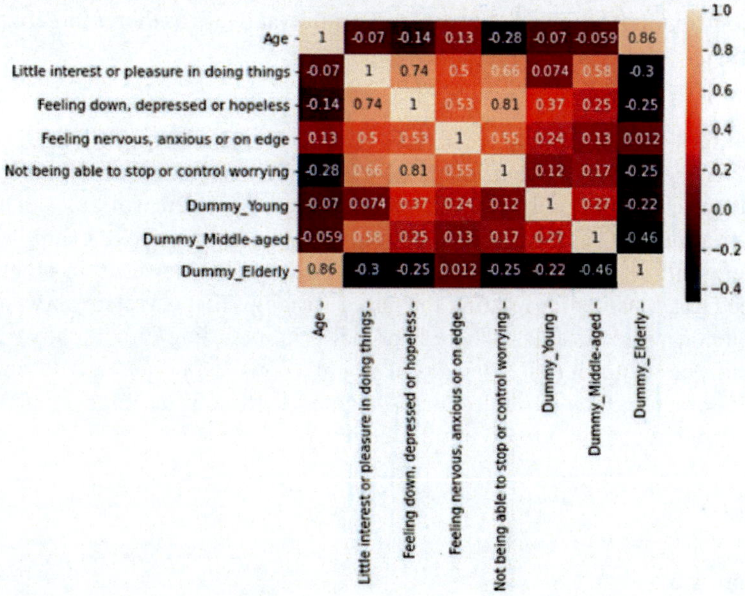


```
<Figure size 432x288 with 0 Axes>
```

Fig. 3.32 Code to display the heat map plus export the correlation matrix to a CSV file (saved on your local computer)

The answer is on the next page.

- **Answer:** Add '`import seaborn as sns`' statement.

Alternatively, you can add these two codes to save the figure and export the correlation output in CSV format to the cell above. The output will be the same. Then it would look as presented in Fig. 3.32.

In the example above, we did not find any pairs of variables with high correlation in this sample dataset. But, of course, there will be dozens of more variables in the real-world dataset, and chances are you will find highly correlated pairs of variables. So it would be safe to remove one of them because, in terms of data, they are

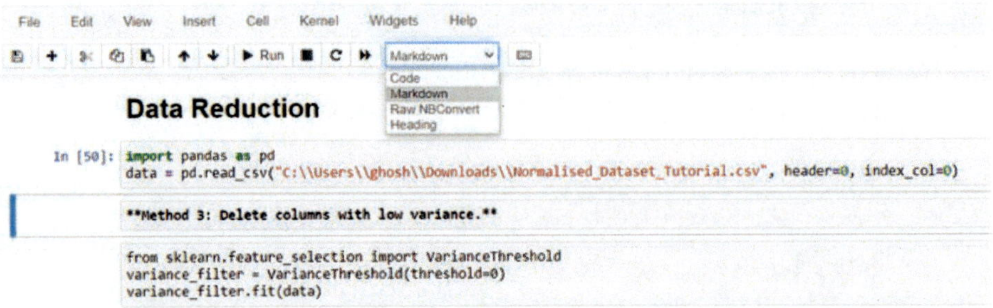

Fig. 3.33 Creating subheadings inside Jupyter Notebook (presentation skill)

not contributing anything new to the output and are rather more like duplicate columns.

Bonus 1
I have mentioned this before but let me repeat it here. These methods or strategies reduce the number of independent/predictor/input variables. Therefore, even though I have correlated all the variables in the above example, one of them will definitely be the target or predicted variable. Still, in practice, you need not consider correlating the target variable. Just correlate between all the pairs of input variables. There is no harm in including the output variable for your personal reference because it might allow you to get an overview of the dataset and which variables are associated with your target, even if that's not what we are aiming to do in this section.

Bonus 2
If you wonder how I created the text subheadings inside the Jupyter Notebook, see Fig. 3.33. All you need to do is click on the cell and change the default from "Code" to "Markdown" from the options shown. For example, if you want the text to be **BOLD**, put ** before and after the text content and then hit RUN. Likewise, if you wish to create a heading like "Data Reduction", simply click on Heading instead of Markdown.

3.4.5 Strategy 5

Filter Out the Important Variables with Feature Selection Using Random Forest Algorithm This technique automatically assigns a score to the independent variables based on their usefulness in predicting the dependent (target) variable. This is an important technique to learn for a psychological scientist. Feature selection not only helps in dimension reduction but also to explain the machine learning model. Therefore, we will dedicate an entire lesson to that (page: 130).

Bonus 1
Supervised and unsupervised machine learning are two types of algorithms. You will learn more about them in the next chapter, but for now, know that a supervised learning algorithm is used to classify data or predict outcomes accurately, while unsupervised learning is used to discover hidden patterns from the dataset.

Bonus 2
A word of caution, we cannot make feature selection for unsupervised machine learning because there is no target or dependent variable. However, it can be useful for classification and regression tasks (supervised learning) when trying to predict a variable from a given set of inputs. We have not introduced you to machine learning yet, so it is completely fine if you do not understand what this means right now. You will soon learn a lot about them in the next chapter.

Bonus 3
If you are still not sold on the importance of dimensionality reduction, then after identifying the less important variables and deleting them from the dataset, run the machine learning model to check for the prediction accuracy. At the same time, run the same machine learning algorithm on the original dataset with all the columns. Then compare the prediction accuracy metrics to see which one is better.

Which Strategy to Use?
Several other approaches include principal component analysis (PCA), forward feature construction and backward feature elimination. Once you feel comfortable with the approaches taught in this book, feel free to experiment with these other methods.

It would be interesting to note that in one of the comparison studies between all these dimensionality reduction methods, it was observed that strategies 2, 3, 4 and 5 (mentioned above) perform better than these three (i.e. PCA, forward feature construction and fackward feature elimination) in terms of reduction ratio, degrading accuracy and speed. The best strategy was feature selection. There is an entire lesson on it (page: 95). So, this section will cover strategies 1, 2, 3 and 4. However, this comparative study was not a rule nor a law. It was not even an academic source but a more informal blog post on the Internet. So, I encourage you to experiment with all these methods on your own data and check which one serves the best for more curious readers.

Bonus

Each of these methods has its pros and cons. So, make sure you check them out to ensure that your research question fits it. For example, some people use principal component analysis (PCA) for dimensionality reduction. But personally, for my research questions, I have not found them a good fit. The reason is PCA transforms the original variables into principal components. These principal components are not as readable and interpretable as original features. In other words, you won't know which predictor or independent variables contributed to your desired output and which did not. While it might benefit some projects focused on prediction accuracies, for psychologists who wish to explain the relationship of the independent variables or predictors with the target or predicted variable, PCA might not be the best choice. PCA might make it difficult to trace back which variables contributed the most to the prediction.

3.5 Final Words

The raw data is often unclean and unfit for data analysis. It requires data cleaning, transformation and reduction to convert into a refined state, enabling data analysis. In this section, we learnt about these. There might be hundreds of variables in your dataset, especially when you use a secondary dataset containing many variables that do not address your research questions. So, it is wise to delete those meaningless or irrelevant variables using your theoretical domain knowledge.

Additionally, there might be theoretically interesting variables for you, but there are 100s of them. Then the question you should be asking is how many of them really contribute to your target (dependent) variable. You will be surprised how often you find only three to four independent variables "substantially" predict the target variable. So, it makes sense to delete the remaining columns/variables from the dataset. There are several approaches you can take. For example, you can sum up different columns into one variable (e.g. symptoms such as low mood and anhedonia can be summed up as depression score); you can delete the variable with too many missing values; you can also delete variables with a low variance; you can delete one among the pairs of highly correlated variables, and you can employ feature selection methods to pick the variables which contribute most towards predicting the target variable. The last one is called feature selection. We will learn about feature selection in a separate lesson because it's one of the most important lessons you need while researching psychological sciences.

References

Ayesha, S., Hanif, M., & Talib, R. (2020). Overview and comparative study of dimensionality reduction techniques for high dimensionality data. *Information Fusion, 59*, 44–58. https://doi.org/10.1016/j.inffus.2020.01.005

Johnston, R., Jones, K., & Manley, D. (2017). Confounding and collinearity in regression analysis: A cautionary tale and an alternative procedure, illustrated by studies of British voting behaviour. *Quality & Quantity, 52*(4), 1957–1976. https://doi.org/10.1007/s11135-017-0584-6

Theodoridis, S., Koutroumbas, K., & Theodoridis, S. (2008). *Pattern recognition* (4th ed.). Academic Press.

Trunk, G. (1979). A problem of dimensionality: A simple example. *IEEE Transactions on Pattern Analysis and Machine Intelligence, 1*(3), 306–307. https://doi.org/10.1109/tpami.1979.4766926

Chapter 4
Machine Learning

Learning Outcome After reading this book chapter, students will be able to analyse *their data using machine learning models.* Here's what you will learn in this chapter: We will start with an overview of machine learning, especially in psychological research. It should make it clear that there are two broad types of machine learning for data analysis in psychological science: supervised ("How well does x predict y?") and unsupervised (to discover meaningful structure in data). Starting with the supervised models, we will learn how to prevent data leakage and solve classification problems. After that, we will learn how to estimate confidence intervals for your machine learning models. There are several machine learning models, but which one to pick? So, the following lesson will help you learn how to run multiple machine learning models on the same data simultaneously and pick the best-performing model. You might wish to improve the performance of the best-performing model and learn how to do hyperparameter tuning to optimise its performance. Until this point, we would demonstrate these different possibilities using one type of supervised machine learning: classifier as the model (used to predict binary/multi-class categorical variables such as YES/NO, TRUE/FALSE). Another supervised machine learning model type is called regressors (used to predict continuous variables such as age or salary). You can implement the same lessons and their respective codes from classifiers with little modifications to solve regression problems. For example, although taught in relation to a classifier, the lesson on the confidence interval can also be transferred to a regressor problem. Next, we will move to regression problems and see how to implement machine learning regressors. These models (classifiers and regressors) generate predictions based on the predictor variables. Still, they do not expose which predictors were more important than others, which might aid in a psychologist's model building/explaining. Therefore, the next lesson we will learn is how to find the most important predictor variables using feature importance functionality. Then, we would cover the first type of machine learning for data analysis, which is supervised machine learning. Supervised machine learning models are meant for building predictive tools or

C. Ghosh, *Data Analysis with Machine Learning for Psychologists*,
https://doi.org/10.1007/978-3-031-14634-3_4

explaining psychological models. There is still another type of machine learning left. That is unsupervised machine learning, which you can use for exploratory investigations. So, then we will learn how to do clustering using hierarchical clustering and K-means clustering. Personality researchers often use dimensionality reduction techniques such as factor analysis. So, we will learn principal component analysis, a type of factor analysis. Finally, there are times when you might be interested in exploring the rules of association between different variables. For that, we will learn rule mining.

4.1 Introduction

Psychologists are curious about many things, and our curiosity leads us to write research questions. Of course, now, as with life, there is more than one path to find the answers. But broadly, such explorations consist of collecting the information (a.k.a. data) from the concerned group of people and then analysing it to find the answers.

Depending on the type of data and the questions we are asking, it is important to pick the right approach to analyse the data, bringing us to the topic of data analysis. A psychology student is heavily trained in statistics for data analysis by their universities. At the time of writing this book, machine learning is quite new to the psychology departments, even in the most advanced schools in the world. So, before getting into the advanced topics of machine learning in this chapter, let me introduce you to some of the basic concepts by comparing machine learning with statistics. That will not only clarify how machine learning is different from statistics but also build up on your existing knowledge, and it will feel more relatable as we proceed.

The Standard Statistical Approach to Data Analysis for Quantitative Psychologists

Researchers start by formulating research question(s). If the person uses inferential statistical analyses, they state the relevant null and alternative hypotheses. Then state the given (probability of a type I error), estimate the degrees of freedom and draw a conclusion based on the calculated test statistic (e.g. Z-tests, t-test, Chi-squared tests and analysis of variance). When the test statistic is in the region of rejection, then we reject the null hypothesis. Still, suppose the test statistic is not in the region of rejection. In that case, we accept the null hypothesis and state the conclusion accordingly. Therefore, the statistical model allows us to test the plausibility of the hypothesis by computing a quantitative measure of confidence that a relationship describes a "true" effect and is unlikely to result from noise or luck (Bzdok et al., 2018). We use statistical significance testing to estimate the probability that a relationship observed in the data occurred only by chance (or due to some factor of interest). You might also have used a combination of p-value testing, effect size measurement and power analysis techniques. This is the classical inferential paradigm. This approach ultimately aimed to generalise from a sample of individuals to a hypothesised population (Bzdok et al., 2018).

A Different Approach to Analysing Data

Statistics are not the only approach to analysing data. You can also use machine learning. Both machine learning and statistics are useful for different purposes, and often they have blurry boundaries between them. There are arguments about how clear the boundaries are better than machine learning and statistics – some even go to the extent of stating that these two are almost synonymous. I have heard arguments that machine learning is just the application of statistics performed on larger datasets to make predictions. But then there are also arguments that statistical tests and machine learning models are substantially different and help us answer different lines of questions (e.g. inference vs prediction). I am inclined to believe the latter with certain reservations. After all, there are also different university courses, research departments and even different job roles for statisticians and machine learning practitioners.

For starters, to keep it simple, remember that:

> Statistics draws population inferences from a sample, and machine learning finds generalizable predictive patterns.
> – Bzdok et al. (2018)

A Brief Overview of Machine learning There are different machine learning models for solving different kinds of problems. For example, supervised machine learning just means you start off with a training set that primes the model (e.g. a classifier or regressor). Unsupervised machine learning like clustering and topic modelling don't use a training set – the system learns how to group different topics or classes on its own, usually via large-scale iteration over the data to detect patterns. Before going in depth on them, let us learn more about machine learning as an alternative (and often complementary) paradigm to analyse the data compared to conventional inferential statistics. But why should we use this another paradigm?

There are different trains of thought:

1. **Machine learning and statistics address different research questions:** There are many types of research questions. For example, at times, we might want to know how psychological processes work (i.e. draw inferences about the relationships between variables). Then there are times when we are interested to know what will happen next (i.e. make the most accurate predictions possible). Inferential statistics can address the first goal. In contrast, the second goal, which aims to forecast unobserved outcomes or future behaviour, is better suited for machine learning to answer. This also indicates two different cultures of data analysis: drawing inferences using hypothesis testing (as we do in statistics) and making predictions using binary classification (as we do in machine learning), respectively (Li & Tong, 2020). But machine learning is more than just predict-

ing the future. Machine learning techniques can also add value to the analysis here by way of detecting patterns or from weights of corresponding features, which tells you "why" are you seeing a specific output and what contributes to it. And that's just one type of machine learning technique called supervised. Unsupervised machine learning techniques can help you find patterns or clusters in your data. Based on what the researcher wants to achieve, they can choose either of them for their work. Still, there might be other research questions that are not answered by either.

2. **Statistics is inferior to machine learning** (at times): The line between statistics and machine learning is often blurry, but there are places where they converge. For example, both are used for prediction. Additionally, some studies demonstrate that machine learning outperforms statistics (e.g. Barnett-Itzhaki et al., 2020). Of course, this might not be true all the time. But I have found this to be the case often. One reason complicated models outperform statistical ones is due to increased complexity, which results in lower bias, but slightly higher variance. Statistical models were designed when computing power was low and you needed an estimate. So, they work very well where you have very fewer data available.
3. On the contrary, machine learning models such as random forest fails here, for it is a "data-hungry" model (i.e. it requires a lot of data to function). So, there are times when simpler models such as linear regression, and logistic regression, have been known to outperform more complex models in practical applications in the industry. So, statistics and machine learning can outperform each other depending on the data and task. Therefore, it is safer to suggest that both machine learning and statistics are essential. Both complement each other, and none is weaker.
4. Another point is that several studies that used inferential statistics have been questioned on issues with replication and reproducibility. On the other hand, machine learning-based analysis has been argued to both maximise accuracy and minimise replicability issues (Orrù et al., 2020). Plus, the validity of inferences drawn from statistical test results relies heavily on how well the data met the associated assumptions (Nimon, 2012). On the contrary, as Orrù et al. (2020) argue, "Machine Learning makes minimal assumptions about the data-generating systems...and in the presence of complicated nonlinear interactions". That said, even while doing data analysis using machine learning models, if your data has varying scales, then it is recommended that you do *normalisation* (MinMaxScaler) or "standardisation" to rescale the different variables to comparable scales having an equal range and/or variance. Otherwise, some of the algorithms may put more weight on the variables with a large range. For example, the variable "salary" will be given more weightage than the variable "age" because salary figures (e.g. £30,000) are larger than the data values of age (e.g. 60 years). We know that both the independent variables, salary and age, are equally important for us to consider and that they are apples and oranges so a £30,000 salary cannot be

equated with 60 years of age. So, we do such rescaling (esp. normalisation), even when the machine learning algorithm you are using does not make assumptions about the distribution of your data, to transform the data to comparable scales. Which one to use, you ask? If you are concerned with outliers affecting your data, then go for standardisation. But *normalisation is* preferred over *standardisation when* the *data doesn't* follow a normal *distribution*. It is a good idea to standardise or normalise the data if different.

5. Machine learning techniques do not make a priori hypotheses about the importance of individual features (which we call "variables" in statistics). Instead, the most informative features can be automatically determined from a potentially large and diverse set of variables and patterns, thereby uncovering previously unknown links between the target and input variables (for review, Hoppe et al., 2018).
6. There are also differences in **language use**. For example, a target is called a "dependent variable" in statistics. Still, the target is called a "label" in classification machine learning (i.e. if it is categorical in nature, like 0 or 1). Likewise, what you call "variable" in statistics is termed a "Feature in machine learning. Being a psychologist, chances are you are more familiar with the language and perspectives of statistics. So, to avoid confusion, I will use the word "target" or output variable to refer to the dependent variable in this book.
7. Finally, there is a difference in the pay scales for the job too. To my knowledge, in most places, machine learning practitioners get paid more than statisticians at many places. But this should not bother you because you are neither applying for the statistician nor a machine learning practitioner post. It is important to position yourself right: You are a psychologist who needs to implement machine learning for data analysis in their academic research project. This book should be enough for you to fulfil most of your needs. In case you want to apply for the post of machine learning at tech giants, this book is the first step for you. You will need advanced coursework.

A while back, we learnt about the two different cultures in data analysis: inference and prediction. Now let us take a one step deeper. Making binary decisions is one of the most common tasks we perform in our daily lives, such as whether to carry an umbrella to work today or offer a particular medicine to a patient and so forth. There is no grey area here. It's pure black and white decision-making. In research, such a categorical research question decides whether a medicine/therapy is an effective intervention for disease X (inferential question) and whether a person has disease X (prediction question). In the first case, the aim was to infer an unknown truth from observed data. In contrast, a prediction task aims to predict an unobserved property of an instance based on historical records – just as a clinical psychologist tries to make binary decisions based on the patient's symptoms (input features). Their decision rules are learnt from previous patients' diagnostic data and the scientific literature.

Similarities Between Statistics and Machine Learning

Both statistics and machine learning are methods to analyse data. Both share terms like regression. The line between statistics and machine learning can be blurry at times. There are overlaps. This book will point to as many of those differences as possible. In this book, we will learn about many machine learning algorithms. But there is something I must clarify at this stage. A lot of different algorithms share a common underlying principle. What this means is best exemplified using the random forest algorithm. Random forest can be used as a regressor as well as a classifier (Liu et al., 2012). Plus, random forest can also be used as a clustering algorithm (e.g. Shi et al., 2004). Therefore, random forest is both a supervised and unsupervised algorithm. For the purpose of this book, we will restrict the use of each algorithm for one purpose. So, if we use random forest for classification tasks, we will not use it for regression or clustering. That way, we will get the opportunity to explore and compare several unique algorithms. But you should not assume that since we used the random forest for classification, it cannot be used for other tasks. After completing the book, I recommend you search the Internet to check the algorithms and how many more types of tasks they can handle.

For now, let's keep it simple. Finally, you will learn different "concepts" under classifier and regressor, such as choosing the best performing model, feature importance and so forth. Note that these concepts under classifier are also same for regressor. Therefore, the codes will also be more or less the same. All you need to do is tweak my regressor codes a little bit to turn them into a classifier code and vice versa. For example, `RandomForestRegressor` will become `RandomForestClassifier`. You might also need to change the evaluation metrics and target variables. But most part of the rest of the code will remain the same. There might be a little modification required. But in case it throws up an error, you know what to do from the lesson on error resolution (Page: 51). Copy and paste the error message to your web browser and find the solution that fits the best for you.

All in all, the difference between statistics and machine learning is often blurry. For example, you will often find methods (e.g. linear regression, logistic regression, hierarchical clustering and principal components analysis) that were originally developed in statistics and are included in the machine learning packages and libraries. Therefore, at this beginner stage, you will save yourself from much unnecessary confusion if you don't bother about whether you should classify a particular model as a statistics or machine learning model. Most often, the answer to the choice of models or approaches lies in the research question.

You are here to learn machine learning in the shortest time possible. So, this chapter will primarily focus on achieving that goal. It makes little sense to invest much time to tease apart the difference between statistics and machine learning, and when they overlap, which one is superior. There are conflicting arguments on both ends with no clear verdict. So, let us jump straight into what really matters.

Machine Learning in Daily Life

We use machine learning algorithms in our daily life today – even without realising it. For example, every time we search on the Internet or get product recommendations, translation services, speech recognition services and self-driving cars – we are using machine learning in practice.

But to a quantitative psychologist, machine learning holds different applications. For example, it can help

- Find the important predictors of a particular target variable (e.g. suicidal behaviour, low mood or treatment response) – supervised machine learning.
- Group individuals or symptoms into distinct groups that are homogenous within themselves – unsupervised machine learning.

An Emerging Trend in Mental Health Prediction Literature Psychological experiment results have been so far largely analysed by orthodox p-based statistical inference and, more currently, by effect size. The replicability crisis of statistical inference in psychology research paved the path for machine learning (ML) methods. Comparatively, machine learning methods are argued to be somewhat superior to statistics in developing models that can generalise/replicate fresh/unseen new data. Additionally, machine learning methods maximise accuracy and minimise replicability issues (for review, Orrù et al., 2020). Therefore, it can be argued that machine learning methods complement more traditional statistical inference techniques.

An increasing body of studies is adopting machine learning to study psychopathology to maximise accuracy and minimise replicability issues. Examples of using machine learning to study psychopathology include the following: differentiating anxiety and depression based on cognitive-emotional biases (Richter et al., 2020), the search for the biomarker of specific disorder vulnerability (Sato et al., 2015), risk prediction models for psychopathology (Meehan et al., 2020) and important variables found for anxiety, stress and depression (Priya et al., 2020), among several others. Further attempts to use machine learning for predicting behaviour, cognitive states and personality have been highlighted in a few studies (e.g. Bixler & D'Mello, 2015, Bulling & Zander, 2014, Henderson et al., 2013, Hoppe et al., 2018).

Imagine that I am teaching you to distinguish patient profiles based on whether their tumour is cancerous or non-cancerous using their medical reports. Let's say I have 100 patients' medical reports in total. So, I decided to use 80 of those to teach you. For simplicity's sake, assume that 40 medical reports suggested cancerous tumours, and the remaining 40 had examples of the non-cancerous tumour. Then once I have taught you the lesson, the next step is to test your knowledge. So, I will show you the rest of the 20 medical reports and ask you to evaluate whether it is a cancerous or non-cancerous tumour. Of course, there will be times when you will be correct, but then there will be times when you will be incorrect. Like you, I will probably teach nine other students using the same medical reports on the same task. Based on how many times you were correct or incorrect, I will pick the best student for placing them in some hospital placement. This is exactly what we are doing with machine learning models.

The way machine learning works is that it attempts to learn the patterns from the historical data of several instances (just as a humans learn through our experiences). Then, based on that, it attempts to predict a new instance.

Now time to take the next step. So far, we have discussed machine learning as a prediction tool. In addition, you can also use machine learning for grouping or clustering similar entities. These entities can be anything. Recently, I attempted to group patients based on their symptoms, for example. While the ML models we use for prediction are called supervised machine learning models, and those which help us cluster entities based on their similarities are called unsupervised machine learning. Therefore, there are two broad machine learning paradigms: supervised and unsupervised. Refer to Fig. 4.1. There is a third type called reinforcement learning, which is interesting because it is inspired by behaviourists (remember B.F. Skinner and positive/negative reinforcements?). It is often used to create robotic agents. To my knowledge, reinforcement learning has not been used to analyse data. We won't cover it in this book because you will often use supervised and unsupervised machine learning for data analysis in psychological sciences.

Returning back to the two types of machine learning, we use supervised machine learning for *prediction tasks* and unsupervised for *grouping tasks*.

1. Sometimes the dataset has clear-cut independent or predictor variables and dependent or target variables. You need to build a predictor model to predict the dependent variable from the corresponding independent variables. That is the job of **supervised machine learning**. It's similar to what we do in statistics with simple linear regression. There are two types of supervised machine learning:

    - **Regression:** It is a regression task when the dependent variable is continuous data, for example, predicting *creativity test scores* of young adults (output) from their age, gender and scores on emotional stability, openness and extraversion scales (input).
    - **Classification:** We call it a classification task when the dependent variable is categorical data. For example, when we predict whether a person will adopt recycling behaviour, YES or NO, from their age, gender, education level and

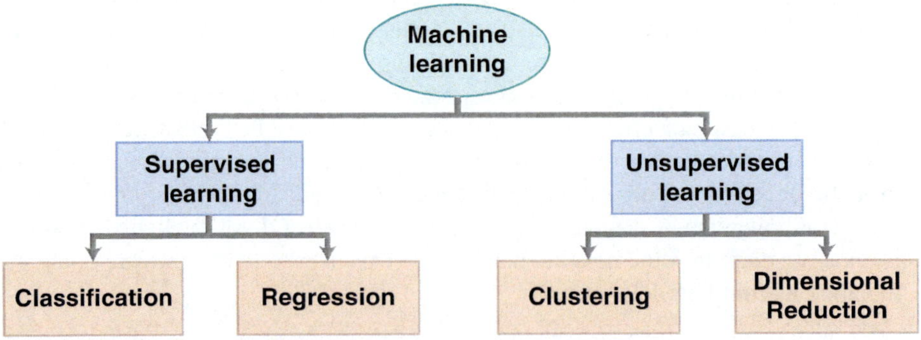

Fig. 4.1 Different types of machine learning algorithms and their application domains

score on the openness scale. Although I have demonstrated binary data for classification in this book, I must notify you that there can be more than two categories in classification tasks: high, moderate and low.

2. Then there are times when the dependent variable is not specified. In such cases, usually, the goal is to explore the data and find some structure within. This is when we apply **unsupervised machine learning** algorithms. There are two types:
   - Clustering: Examples would include clustering groups of patients admitted to a hospital based on their symptoms and residential address. There are two types of clustering:
     - **Hierarchical:** It produces a hierarchy of clusters by either merging smaller clusters into larger ones or dividing larger clusters into smaller ones. It requires a similarity measure.

   1. Agglomerative clustering strategy (bottom-up approach): merging clusters into larger ones
   2. Divisive clustering strategy (top-down approach): splitting clusters into smaller ones

     - **Partitioning:** It generates different partitions and then assesses them by some criterion. It requires you to specify the number of clusters (at the start) and the initial centres.

   Dimensionality Reduction**:** A host of problems emerge as the number of variables increases. For example, the model accuracy decreases. Additionally, algorithms train slower, and more storage space is required. Furthermore, the model is being trained on several redundant variables and noise. So, it is recommended that we reduce the number of variables to a minimum. This is true for researchers using secondary datasets where you use someone else's dataset to analyse. That dataset might have hundreds of variables in the dataset. Out of them, probably you need just five or ten variables (for example). Therefore, if you are trying to build a supervised machine learning model on a dataset with many variables, it may be useful to reduce it with an unsupervised step before supervised steps.

   There are two approaches:

Feature Selection
The simplest approach is identifying and retaining relevant variables for your study and deleting every other variable (representing columns). You can use methods like correlation and variance threshold (as taught on page: 78, 79). You can even use machine learning-based methods such as using "feature importance" and SHAP values (you will learn later on page: 134).

Feature Engineering
In addition to feature selection, there is feature engineering. Here we generate new variables from existing variables by applying some transformation or performing some operation on them. There are multiple ways of doing it. For example, one common way psychological scientists do this is to sum up multiple variables (e.g.

item-level scores from a particular questionnaire) to create a single variable (e.g. total depression score). Some other people use factor analysis, principal component analysis, t-distributed stochastic neighbour embedding (t-SNE), linear discriminant analysis and other methods. You might have heard about them in statistics even if you have not used them. Without getting into the mud of whether they are statistics or machine learning, let us focus on these methods that can allow you to reduce the dimensionality:

- Principal component analysis (PCA, from `sklearn.decomposition import PCA`).
- Random projection methods: Three examples of them are as follows with how to import them from sklearn machine learning library:
  a. The Johnson-Lindenstrauss lemma (`from sklearn.random_projection import johnson_lindenstrauss_min_dim`)
  b. Gaussian random projection (`from sklearn import random_projection.GaussianRandomProjection`)
  c. Sparse random projection (`from sklearn import random_projection.SparseRandomProjection()`)
- Feature agglomeration (this includes applying hierarchical clustering to group together variables that behave similarly). We will learn about hierarchical clustering under the clustering lesson (page: 136). Note that if the variables have very different scaling, then `cluster.FeatureAgglomeration` may not be able to capture the links between related variables. In such cases, use `preprocessing.StandardScaler` (page: 73) beforehand.

If you ask me, then I have never used PCA, random projection or feature agglomeration for dimensionality reduction during my psychological research. The argument is that it reduces the original variables into a new set of uncorrelated attributes. To realise the problem with this approach, check out the BEFORE and AFTER doing PCA (refer to Figs. 4.2 and 4.3):

Before PCA

After PCA

When doing research in psychology, I imagine that you are trying to understand a phenomenon such as mental illness or social behaviour (for example). When you do PCA, the variables get fused together, and you get the output as principal components A and B. Now while you have a lesser number of columns now (solving the problem with dimensionality), what you lost was the information about specific variables. Even if principal components A and B predict the target (e.g. suicide attempt) with 99% accuracy, what can you infer about the phenomena or the predictors? We do not know which symptoms predicted the suicidal attempt. Therefore, in my experience, I have found that using methods such as PCA is unhelpful in understanding the relationships between independent and dependent variables. That's what most of our research is about, isn't it?

However, these methods, such as PCA, do have merit for other kinds of projects. Imagine that there might be machine learning practitioners in the industry who are

Little interest or pleasure in doing things	Feeling down, depressed or hopeless	Feeling nervous, anxious or on edge	Not being able to stop or control worrying
1.000000	1.000000	1.000000	1.000000
0.000000	0.000000	0.000000	0.000000
0.333333	0.333333	0.333333	0.666667
0.301587	1.000000	1.000000	1.000000
0.333333	0.333333	0.666667	0.449275

Fig. 4.2 The dataset before performing PCA

	principal component A	principal component B
0	1.244072	-0.027247
1	-0.753465	-0.108243
2	0.084739	-0.150064
3	0.902447	0.177677
4	0.126968	0.189195

Fig. 4.3 The dataset after performing PCA

using machine learning to build a better recommender system for a video streaming or e-commerce app. For such use cases, they are not interested in learning "what predicts a particular target?". Instead, their focus is on achieving the model that generates an output of the highest accuracy.

There might be use cases in psychology as well. For example, in personality research, PCA is the preferred type of factor analysis (Goekoop et al., 2012). Then there are machine learning studies which used PCA to reduce the dimensions or variables in the dataset. You might find it interesting that although I have introduced PCA as an unsupervised machine learning, the same is also seen as a classical statistical method. Such labels or groupings, in my opinion, do not matter as much as knowing that PCA can be useful in specific studies and how to do PCA. So, whether you call it statistics or machine learning, I have demonstrated how to do PCA on page 148 because you might find it useful.

To sum up, supervised machine learning is used when you have something to predict. On the other hand, we use unsupervised machine learning when the goal is to cluster the data observations (e.g. group people based on their similarity in characteristics) or reduce the dimensions.

Homogeneity of Sample

Since we psychologists often work with human data, I cannot stress enough that your sample should have a homogenous group whether you are using supervised or unsupervised machine learning models. Suppose you are studying suicidal

Table 4.1 An example of demographic grouping

Gender	Male	Female
Age	Young	Young
	Middle age	Middle age
	Elderly	Elderly

behaviour (for example) and wish to understand which of the depressive symptoms are predictors of it. Let's say, and literature shows that age and gender are important predictors of suicidal behaviour (refer to Table 4.1). Since you are not studying age and gender, such potential extraneous variables might confound your results. Therefore, you must take all possible measures to prevent its influence. One way is to hold them constant by taking a homogenous group of people. In one of my recent studies (for example), I divided my sample into six subgroups (as in the table below). I built a machine learning model on each of them. I found the relative importance of each predictor variable on the target variable. Subsequently, I compared the six findings for similarities and dissimilarities to show how people of different ages and gender vary in terms of the predictors of the target behaviour. This is similar to a study on finding the determinants of happiness. For example, one might find that watching movies with friends is the most important predictor of happiness for a young male. Still, the most important determinant of happiness might be knitting sweaters for an elderly female.

Application of Machine Learning by Psychologists

There are multiple routes to analyse data. In conventional mental health literature, statistics has been the key route. But recently, there have been rapid developments in analysing mental health data using machine learning.

Machine learning is a method that can find patterns from the data and can then use that uncovered patterns to predict new data. For example, behavioural and Social Scientists popularly used ordinary least-squares (OLS) regression techniques to find data patterns. Then, when the goal is to test the hypothesis, such patterns (usually regression coefficient) were examined for statistical significance to determine the effect of predictors on the outcome variable.

Developing and testing a hypothesis is an important route of conducting research. But some researchers are more interested in making accurate predictions (e.g. will a patient with symptoms X experience Y?). For example, inferential tests might reveal a significant correlation between X and Y. However, it does not tell us how accurately X predicts Y. Is the model accurate enough to implement interventions or policies?

Statistics can again help; many behavioural and social scientists use linear regression to achieve such goals (e.g. Hassan et al., 2016; Rimfeld et al., 2016). But only this time, the regression model coefficient is no longer the primary metric (as it was during hypothesis testing). Instead, the predicted accuracy is the main metric. Along with statistics, machine learning has another approach to predicting the future.

Machine learning, unlike statistics, divides the dataset into two parts: the training dataset (80% of the data) and the test dataset (20% of the data). It learns the patterns from the training dataset. It then uses the unseen test data to evaluate its learning accuracy (i.e. how close predictions are to observed values). The use of test data (foreign data or unseen by the model) to predict and report the accuracy scores is called out-of-sample prediction accuracy.

The functioning mentioned above of machine learning is similar to how a human learns. A student (machine learning model) learns from different books (training set) and then appears for an examination (question paper is like test set) to get grades (performance evaluation criteria). Of course, some students will solve the question paper more accurately than the rest to pick the best performer for the job (e.g. deployment in a web application).

Machine learning is chosen over statistics, typically used when the focus is on prediction accuracy, when the study is interested in predicting new (unseen) samples (such as the development of a predictive software) and when the study wants to manually tune the model to better fit the problem at hand. These are not possible with statistics.

Studies using supervised machine learning models fall in mainly two categories: First is to generate an algorithm capable of predicting the future value of the variable of interest (e.g. risk of suicide, Roy et al., 2020). Second, examining predictors of the variable of interest that depend on its nature might be critical in intervention and prevention efforts (e.g. predictors of suicidal ideation, Rozek et al., 2020). So, we will use regressors or classifier models, depending on whether the outcome variable is continuous or categorical (e.g. Yes/No, True/False).

Supervised machine learning is "supervised" because the human researcher supervises which one is the predicted variable and which one is the predictor variable. But there is another vital application of machine learning: unsupervised machine learning, which allows researchers to detect theoretically meaningful patterns in the data (Jordan & Mitchell, 2015) without human supervision (or the requirement for specific predicted/predictor variables). This is usually done using clustering methods. Unfortunately, there are few studies in psychological sciences that use unsupervised machine learning or clustering models. But such clustering approaches can be very useful when the research goal is to group people (or any interest entities) in terms of similarity. An example might be patients with similar symptoms, which might be clustered or grouped together using unsupervised machine learning.

Another infamous example of applying clustering methods is the Cambridge Analytica case, which made machine learning even more popular in psychological science. A brief history of what seemed to happen was Cambridge Analytica got access to millions of American Facebook (FB) users and created their psychographic profiles using machine learning based on their FB behaviours (i.e. Likes). The psychographic profiling informed Cambridge Analytica of the FB users who were "open to suggestion". Then, using FB advertisement, those specific

Table 4.2 Differences between statistics and machine learning

Traditional statistics modelling	Machine learning modelling
Prediction is constrained by strong assumptions about certain characteristics of the data.	Not a lot of emphasis on assumptions
Focus on inference (e.g. inferential statistics)	Focus on prediction (e.g. supervised machine learning) or grouping or clustering (e.g. unsupervised machine learning)
The relationships between the covariates and outcome are linear relationships (Jacobucci & Grimm, 2020)	The relationships between the covariates and outcome are non-linear (Jacobucci & Grimm, 2020)
Hypothesis testing	Predictive accuracy

persuadable voters were shown custom-made campaign ads favouring candidate Donald Trump (the US presidential election, 2016).

The above-reported example of the application of machine learning was outside the academic arena. However, behavioural and social scientists are increasingly using machine learning either as an alternative or to complement traditional statistics within the academic domain. It has been argued that psychological experiments with machine learning-based analysis will have higher accuracy and lower replicability issues than statistical inference (Orrù et al., 2020). Furthermore, machine learning models such as the random forest have consistently demonstrated their superiority over traditional statistical models such as logistic and linear regression (e.g. Tasmim et al., 2020). Usually, the usage of machine learning is recommended when the relationships between the covariates and outcome are nonlinear (Jacobucci & Grimm, 2020).

Although the boundaries between machine learning and statistics are fuzzy, some marked differences are enlisted in Table 4.2.

As reviewed in Rosenbusch et al. (2021), experimental psychologists can use machine learning techniques to explore differences between experimental conditions (Koul et al., 2018). In addition, personality researchers can use them to assess the validity of psychological constructs (Bleidorn & Hopwood, 2019). Thus, while we concentrated on prediction accuracy, machine learning also offers much to theory-driven research.

The higher performance of machine learning models is compromised if the quality of data is poor. It has been demonstrated that predictor variables' measurement quality (error) can negatively influence prediction, particularly changes in the explained variance (Jacobucci & Grimm, 2020). Therefore, to get optimum performance from the analytic data models, it is important to ensure that we feed in higher-quality data and take steps to ensure it is the case.

4.2 Classification

4.2.1 Getting Started with Supervised Machine Learning

As said before in the "Types of Research" section, most psychological studies today focus on explaining behaviour. You can implement machine learning techniques to assist the conventional statistical modelling and conduct better explanatory studies. If you are feeling busy then re-read at least the part that discusses explanatory research. In general, explanatory studies are not without concerns. They cannot typically predict future experiences or behaviour with reasonable accuracy. That's where predictive research comes into play. Psychologists can use machine learning techniques to engage in such predictive research and predict future observations as accurately as possible (i.e. minimise prediction error).

Machine learning is versatile. It's not just for predicting future behaviour or experience. But once you have a predictive model, machine learning allows you to "explain" it. We will learn that in the lesson for "feature importance" later in this book. For now, know that feature importance refers to the group of techniques that are used to assign scores to predictor variables of a predictive model that indicates the relative importance of each predictor when making a prediction. This allows us to use machine learning and make academic contributions to the literature. For example, it can help you identify which variables are important predictors of beneficial outcomes of CBT. We might find that age is an important contributor or determinant of the extent to which someone might benefit from CBT. On further investigation, let's say we found that younger people are benefited from CBT more than the elderly. This kind of insight can help us improve the mental healthcare service. We will learn about feature importance later.

In the current lesson, we will learn how to solve predictive problems using supervised machine learning. For example, you might be interested in predicting the treatment outcome of CBT (e.g. the decline in symptom frequency) and which individual patients will benefit from the CBT. That is the power of predictive research. Then, you can develop a predictive tool or computer application using the results and distribute them among the clinicians for use.

There are four simple steps:

1. Collect the data, prepare it and divide it into "training set" and "test set".
2. Select an appropriate algorithm(s) to run on the data depending on the type of research question and nature of the data.
3. Train the chosen algorithm(s) using the training dataset to build a machine learning model.
4. Test the model's performance using the test dataset, which the model has never seen the test data.

Let's dive into an example straight away. This idea is to start with an overview and then go into variations of problems and, in doing so, learn all about machine learning. So far, we have been using a small-sized fictional dataset for your ease of understanding. But the time has come to take the next step and analyse some real-world datasets. The subsequent topics will be taught using real-world data, called the Coronavirus (COVID -19) Psychological Wellbeing Study (Armour et al., 2020). Let's quickly run through the steps of data pre-processing. We have covered this in the last lesson, so I won't be explaining it again. Take a close look and see if you remember or make sense of each of the lines of code.

Step 1: Import (Fig. 4.4)

Step 2: Data pre-processing (Fig. 4.5)

In the code above (Fig. 4.6), the .describe() allows us to get a quick glance into the descriptive statistics of the dataset. From the output of that cell [588], you can see it summarises the central tendency, dispersion and also the shape of a dataset's distribution. Each column's min and max values can also help us spot potential errors. For example, if you found the max value under the variable "age" of a human participant is 450 years, then you know it is potentially a wrong value. Likewise, the `.drop()` has been used to delete or remove any columns you think are irrelevant to the investigation. In this case, let us assume that the "some irrelevant column" represents a duplicate column.

Notice that we used `.get_dummies` in the code given in Fig. 4.7. We are creating dummy variables.

Why Do You Ask?

Sometimes, your dataset will have categorical variables, such as gender. Now there are different types of gender outside male and female, such as transgender, cisgender, non-binary, gender-fluid and so forth. But for the sake of simplicity of learning

```
In [446]: import pandas as pd
          import numpy as np
          from sklearn.model_selection import train_test_split
          from sklearn.linear_model import LogisticRegression
          from sklearn.tree import DecisionTreeClassifier
          from sklearn.preprocessing import MinMaxScaler
          from sklearn.metrics import confusion_matrix
          from sklearn.metrics import accuracy_score
          from sklearn import metrics
          from sklearn.metrics import roc_auc_score
          from yellowbrick.classifier import ClassificationReport
          import seaborn as sns
```

Fig. 4.4 Codes to import the required libraries and modules to perform classification

```
In [447]: data = pd.read_csv("C:\\Users\\ghosh\\Downloads\\Sample_Data_.csv", header=0, index_col=0)
          data.head() #For identifying duplicate columns, and categorical variables
```

Out[447]:

ID	Gender	Age	Feeling nervous, anxious or on edge	Not being able to stop or control worrying	Little interest or pleasure in doing things	Feeling down, depressed, or hopeless	Thoughts that you would be better off dead or of hurting yourself in some way	Some Irrelevant Variable
Respondent_1921	Female	36	3	3	3.0	3.0	1.0	0.0
Respondent_126	Female	40	3	3	2.0	2.0	0.0	0.0
Respondent_1722	Female	30	3	3	2.0	2.0	3.0	1.0
Respondent_1926	Female	46	3	2	3.0	3.0	2.0	0.0
Respondent_1786	Female	39	3	3	3.0	3.0	3.0	0.0

Fig. 4.5 Codes to display the first five rows of the dataset

```
In [588]: data.describe()
Out[588]:
```

	Age	Feeling nervous, anxious or on edge	Not being able to stop or control worrying	Little interest or pleasure in doing things	Feeling down, depressed, or hopeless	Thoughts that you would be better off dead or of hurting yourself in some way	Some Irrelevant Variable
count	1318.000000	1318.000000	1318.000000	1315.000000	1315.000000	1315.000000	1308.000000
mean	36.171472	1.738240	1.355083	1.133080	1.257795	0.373384	0.098624
std	11.886685	0.845145	1.015379	0.957908	0.964991	0.781088	0.298270
min	18.000000	1.000000	0.000000	0.000000	0.000000	0.000000	0.000000
25%	27.000000	1.000000	1.000000	0.000000	1.000000	0.000000	0.000000
50%	34.000000	1.000000	1.000000	1.000000	1.000000	0.000000	0.000000
75%	44.000000	3.000000	2.000000	2.000000	2.000000	0.000000	0.000000
max	76.000000	3.000000	3.000000	3.000000	3.000000	3.000000	1.000000

```
In [589]: data = data.drop(columns =['Some Irrelevant Variable']) #remove any columns of your choice

In [590]: print(data.isnull().sum()) #Display all the missing values for each column or variable

Gender                                                                      0
Age                                                                         0
Feeling nervous, anxious or on edge                                         0
Not being able to stop or control worrying                                  0
Little interest or pleasure in doing things                                 3
Feeling down, depressed, or hopeless                                        3
Thoughts that you would be better off dead or of hurting yourself in some way    3
dtype: int64
```

Fig. 4.6 Codes to describe, remove irrelevant (e.g. duplicate) columns and diagnose all missing values in each column

here, let us stick to two categories: male and female. To us humans, each of these words means something different. But computers are not great at analysing "words" – they are brilliant with numbers, though. So, we convert this categorical variable into a numeric variable by assigning one of the two genders as "1" and another as "0". As you can see in Fig. 4.7, we have assigned "0" to all males and "1" to all females and accordingly created a dummy variable called "Binary_Gender_Male" (from a data-analysis viewpoint, we don't need "Binary_Gender_Female"). The `.get_dummies` converted the categorical variable "gender" into a dummy variable which now can be used to build the prediction model.

Time and again, I have come across several psychology literature studies that have divided participant scores into binary or more than binary categories. Such as "depressed"/"not-depressed", "Highly anxiety"/"Medium anxiety"/"Low anxiety" and so forth. If you want to achieve that, look at Fig. 4.8 (cell [487]).

You can add the following line of code at the end if you want to download the pre-processed data for future reference or use (Fig. 4.9):

Bonus

If you notice the codes in Fig. 4.8, I have added something "to convert a continuous variable into a categorical variable". If you wish to create more than two categories, then it's still the same code. All you need to do is enter the additional category_labels and if it's three, then enter that among bins =. If you wish to categorise the target variable into two labels (high and low) then the code then the code is, as represented in Fig. 4.10.

```
In [451]: data.dropna(inplace = True) #Remove the rows with Missing Values
          print(data.isnull().sum())

          Gender                                                                     0
          Age                                                                        0
          Feeling nervous, anxious or on edge                                        0
          Not being able to stop or control worrying                                 0
          Little interest or pleasure in doing things                                0
          Feeling down, depressed, or hopeless                                       0
          Thoughts that you would be better off dead or of hurting yourself in some way   0
          dtype: int64

In [452]: data = data.drop_duplicates(keep='first') #remove duplicate rows

In [453]: data = pd.get_dummies(data, columns=['Gender'], drop_first=True, prefix='Binary_Gender')
          data.head(5)
Out[453]:
```

ID	Age	Feeling nervous, anxious or on edge	Not being able to stop or control worrying	Little interest or pleasure in doing things	Feeling down, depressed, or hopeless	Thoughts that you would be better off dead or of hurting yourself in some way	Binary_Gender_Male
Respondent_1921	36	3	3	3.0	3.0	1.0	0
Respondent_126	40	3	3	2.0	2.0	0.0	0
Respondent_1722	30	3	3	2.0	2.0	3.0	0
Respondent_1926	46	3	2	3.0	3.0	2.0	0
Respondent_1786	39	3	3	3.0	3.0	3.0	0

Fig. 4.7 Codes to delete rows with missing values, remove duplicate rows and create dummy variables

```
In [487]: #Converting a continuous variable into a binary categorical variable
          category_labels = ['High', 'Low']
          data['Feeling nervous, anxious or on edge_New'] = pd.cut (x = data['Feeling nervous, anxious or on edge'],
                                                                    bins = 2, labels = category_labels)
          data = pd.get_dummies(data, columns=['Feeling nervous, anxious or on edge_New'], drop_first=True,
                                prefix='Binary_Feeling nervous, anxious or on edge_')
          data.head(5)
Out[487]:
```

ID	Age	Feeling nervous, anxious or on edge	Not being able to stop or control worrying	Little interest or pleasure in doing things	Feeling down, depressed, or hopeless	Thoughts that you would be better off dead or of hurting yourself in some way	Binary_Gender_Male	Binary_Feeling nervous, anxious or on edge__Low
Respondent_1921	36	3	3	3.0	3.0	1.0	0	1
Respondent_126	40	3	3	2.0	2.0	0.0	0	1
Respondent_1722	30	3	3	2.0	2.0	3.0	0	1
Respondent_1926	46	3	2	3.0	3.0	2.0	0	1
Respondent_1786	39	3	3	3.0	3.0	3.0	0	1

```
In [488]: data['Binary_Feeling nervous, anxious or on edge__Low'].value_counts()
Out[488]: 0    765
          1    308
          Name: Binary_Feeling nervous, anxious or on edge__Low, dtype: int64
```

Fig. 4.8 Codes to convert a continuous variable into a binary categorical variable and then convert that into a dummy variable

Step 3: Then, we will use machine learning for data analysis. That's step 3, and you will see the codes below. I will demonstrate to you two approaches to do so. The first is the leaky (wrong) approach, and the second is the leak-proof approach.

> **Bonus**
> This leaky vs non-leaky approach bothers only when you are solving classification or regression problems (supervised machine learning models). It does not apply when you are solving clustering or dimensionality reduction problems (unsupervised machine learning models).

```
In [110]: data.to_csv(r'C:\\Users\\ghosh\\Downloads\\Pre-processed dataset.csv')
```

Fig. 4.9 Code to download the pre-processed data

```
In [25]: #Converting a continuous variable into a binary categorical variable
         category_labels = ['High', 'Low']
         data['Feeling nervous, anxious or on edge_New'] = pd.cut (x = data['Feeling nervous, anxious or on edge'],
                                                                  bins = 2, labels = category_labels)
         data = pd.get_dummies(data, columns=['Feeling nervous, anxious or on edge_New'], drop_first=True,
                               prefix='Binary_Feeling nervous, anxious or on edge_')
         data.head(5)
Out[25]:
```

ID	Age	Feeling nervous, anxious or on edge	Not being able to stop or control worrying	Little interest or pleasure in doing things	Feeling down, depressed, or hopeless	Thoughts that you would be better off dead or of hurting yourself in some way	Binary_Gender_Male	Binary_Feeling nervous, anxious or on edge__Low
Respondent_1921	35	3	3	3.0	3.0	1.0	0	1
Respondent_126	40	3	3	2.0	2.0	0.0	0	1
Respondent_1722	30	3	3	2.0	2.0	3.0	0	1
Respondent_1926	45	3	2	3.0	3.0	2.0	0	1
Respondent_1786	39	3	3	3.0	3.0	3.0	0	1

```
In [26]: data['Binary_Feeling nervous, anxious or on edge__Low'].value_counts()
Out[26]: 0    765
         1    308
         Name: Binary_Feeling nervous, anxious or on edge__Low, dtype: int64

In [27]: data = data.drop(columns =['Feeling nervous, anxious or on edge']) #remove duplicate columns
```

Fig. 4.10 Convert a continuous variable into a categorical variable (binary)

```
In [565]: Variable_Scaler = MinMaxScaler()
          scaled_data = pd.DataFrame(Variable_Scaler.fit_transform(data))

          #Print the normalised data
          print (scaled_data.head())

          #Exporting the normalised data to your computer
          scaled_data.to_csv('Scaled_Dataset_Tutorial.csv')

          0         1         2         3         4    5    6
0  0.310345  1.000000  1.000000  1.000000  0.333333  0.0  1.0
1  0.379310  1.000000  0.666667  0.666667  0.000000  0.0  1.0
2  0.206897  1.000000  0.666667  0.666667  1.000000  0.0  1.0
3  0.482759  0.666667  1.000000  1.000000  0.666667  0.0  1.0
4  0.362069  1.000000  1.000000  1.000000  1.000000  0.0  1.0
```

Fig. 4.11 Normalisation/standardisation on the entire dataset (it's a wrong approach)

One of the main areas where psychologists using machine learning go wrong is here. So, I thought in this book I should inform you of this in detail with examples so that you don't repeat the same.

The Leaky Approach (for Those Working with Supervised Models)

As mentioned, the first step is importing all the required Python libraries, modules and packages. Then we did data pre-processing. During this time, some people (mistakenly) perform feature transformation (e.g. normalisation or standardisation) on the entire dataset, as shown in Fig. 4.11.

We will understand why it can be a flawed approach in a short while, but since you might use this in a different context, appropriately focus on the output of the rescaled dataset.

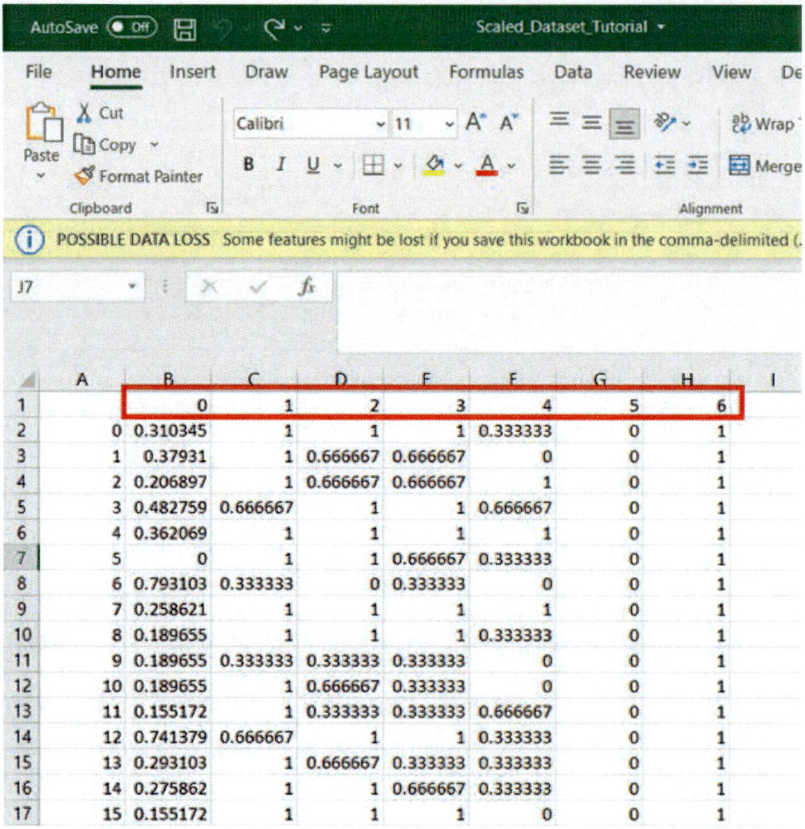

	A	B	C	D	E	F	G	H
1		0	1	2	3	4	5	6
2	0	0.310345	1	1	1	0.333333	0	1
3	1	0.37931	1	0.666667	0.666667	0	0	1
4	2	0.206897	1	0.666667	0.666667	1	0	1
5	3	0.482759	0.666667	1	1	0.666667	0	1
6	4	0.362069	1	1	1	1	0	1
7	5	0	1	1	0.666667	0.333333	0	1
8	6	0.793103	0.333333	0	0.333333	0	0	1
9	7	0.258621	1	1	1	1	0	1
10	8	0.189655	1	1	1	0.333333	0	1
11	9	0.189655	0.333333	0.333333	0.333333	0	0	1
12	10	0.189655	1	0.666667	0.333333	0	0	1
13	11	0.155172	1	0.333333	0.333333	0.666667	0	1
14	12	0.741379	0.666667	1	1	0.333333	0	1
15	13	0.293103	1	0.666667	0.333333	0.333333	0	1
16	14	0.275862	1	1	0.666667	0.333333	0	1
17	15	0.155172	1	1	1	0	0	1

Fig. 4.12 Headers missing from the output of normalisation/standardisation in Python (step 1) – opened in CSV format

It will appear when you open the saved file (.csv), as shown in Fig. 4.12. Note this because even when you make the "correct" approach, the same problem will appear in Python, and you will need to fix it using a simple copy-paste solution.

Notice that the headers went missing. There are several ways to bring back the column headers. The simplest one, in my opinion, is to open the CSV file and copy-paste the header from the original dataset as shown in Fig. 4.13.

So, we will read the above dataset, which has already been cleaned and transformed using the following code in Fig. 4.14.

Bonus

How will you know which packages, modules and libraries are relevant, you ask? Follow through step 1 of each machine learning code script in the upcoming pages and note the common ones (e.g. pandas, NumPy, sklearn and Seaborn). You will see that only a few of them are different for each script. You will also notice that specific entities are being imported from big modules like sklearn, such as confusion_matrix, metrics and so forth.

One other area of difference you will notice in the different code scripts is the choice of models. Suppose you wish to do a classification task with a decision tree. In that case, we will import the decision tree classifier from the sklearn. Likewise, while doing a regression task with a decision tree, we will see the importance of a decision tree regressor instead. Many of these models can be used for classification and regression tasks. Similarly, suppose you intend to use more conventional statistical models such as linear regression model (e.g. ordinary least squares). In that case, you will import a module called "statsmodels".

	A	B	C	D	E	F	G	H	I	J
1	ID	Age	Not being able	Little interest o	Feeling down	Thoughts that	Binary_Gender_	Binary_Feeling	ervous, anxic	
2	0	0.310345	1	1	1	0.33333333	0	1		
3	1	0.37931	1	0.666666667	0.66666667	0	0	1		
4	2	0.206897	1	0.666666667	0.66666667	1	0	1		
5	3	0.482759	0.666666667	1	1	0.66666667	0	1		
6	4	0.362069	1	1	1	1	0	1		
7	5	0	1	1	0.66666667	0.33333333	0	1		
8	6	0.793103	0.333333333	0	0.33333333	0	0	1		
9	7	0.258621	1	1	1	1	0	1		
10	8	0.189655	1	1	1	0.33333333	0	1		
11	9	0.189655	0.333333333	0.333333333	0.33333333	0	0	1		
12	10	0.189655	1	0.666666667	0.33333333	0	0	1		
13	11	0.155172	1	0.333333333	0.33333333	0.66666667	0	1		
14	12	0.741379	0.666666667	1	1	0.33333333	0	1		
15	13	0.293103	1	0.666666667	0.33333333	0.33333333	0	1		
16	14	0.275862	1	1	0.66666667	0.33333333	0	1		
17	15	0.155172	1	1	1	0	0	1		
18	16	0.034483	1	1	1	1	0	1		
19	17	0.413793	1	1	1	1	0	1		
20	18	0.344828	1	1	1	1	0	1		

Fig. 4.13 Headers copied from the raw dataset and then pasted on the standardised or normalised dataset (step 2) – opened in CSV format

```
In [570]: data = pd.read_csv("C:\\Users\\ghosh\\Downloads\\Scaled_Dataset_Tutorial.csv", header=0, index_col=0)
          data.head()
Out[570]:
```

ID	Age	Not being able to stop or control worrying	Little interest or pleasure in doing things	Feeling down, depressed, or hopeless	Thoughts that you would be better off dead or of hurting yourself in some way	Binary_Gender_Male	Binary_Feeling nervous, anxious or on edge__Low
0	0.310345	1.000000	1.000000	1.000000	0.333333	0	1
1	0.379310	1.000000	0.666667	0.666667	0.000000	0	1
2	0.206897	1.000000	0.666667	0.666667	1.000000	0	1
3	0.482759	0.666667	1.000000	1.000000	0.666667	0	1
4	0.362069	1.000000	1.000000	1.000000	1.000000	0	1

Fig. 4.14 Transformed dataset (after going from steps 1 and 2) – opened in Jupyter Notebook

Leaky Approach (Simple and Naïve Yet Flawed) Refer to the codes in Fig. 4.15.

Step 3: Tell Python which variables to consider. This is an opportunity for you to manually NOT include any variable that looks irrelevant to your research question. For example, I have excluded age. As you might remember from the first lessons in this book, always keep the textual content with quotation marks. So never forget to put those quotation marks to specify each variable. After that, specify which variables out of those selected variables (`"feature_names"`) are predictors (or independent) and which one is the target/predicted (dependent). I have selected a binary variable as the target variable, so I expect to predict if a person will attempt suicide based on the four symptoms.

Step 4: We need to divide the dataset into two parts: training and test sets. The idea is simple, and the model will learn the patterns from the training set. Then we will use a test set to evaluate how well the model has learnt the patterns (using accuracy score).

Step 5: We will finally build the decision tree classifier and make it learn from the training dataset (represented by X_train and y_train).

Bonus

This is a different approach from what we would be inquiring about during hypothesis testing. Note these differences as this is an important factor that distinguishes between hypothesis testing (which you have already learnt in stats classes) and binary classification (which I am showing you here). The aim of binary classification (machine learning method) is to try to predict what a particular person will do. For example, suppose a patient comes to a mental healthcare clinic. In that case, you can ask the person to rate their responses to those four symptoms (see the dataset). Once you feed that on your machine learning app, it will tell you whether this particular person is about to attempt suicide. So, if you have already noticed, machine learning (at least binary classification) focuses on learning the patterns from historical data and then using that knowledge to build apps capable of making predictions about individuals. This is something different from what statistical hypothesis testing is capable of doing. Hypothesis testing is meant to address research questions that concern an unobservable population-level property of a feature. For example, suppose your research question asks whether a gene is a biomarker of depression. In that case, it is a hypothesis testing task. On the other hand, if your line of inquiry is to determine whether an individual has depression, you must do binary classification.

As mentioned before, the X represents the independent or predictor variables, while the "y" represents the target or predicted variable. During the test, note that the model won't have access to the y_test. Instead, it will try to predict (y_pred =) whether the person attempted suicide based on the X_test. Then it will compare its

```
In [507]: #Step 3: Assign variables as predictors and target
feature_names = ["Binary_Gender_Male", "Age", "Not being able to stop or control worrying",
                 "Little interest or pleasure in doing things", "Feeling down, depressed, or hopeless"]
X = data[feature_names]      #predictors
y = data["Binary_Feeling nervous, anxious or on edge__Low"] #predicted

#Step 4: Divide the dataset into two parts: training set and test set.
X_train, X_test, y_train, y_test = train_test_split(X, y, test_size=0.30, random_state=100)

#Step 5: Build the model
classifier_model = DecisionTreeClassifier()
classifier_model=classifier_model.fit(X_train, y_train)
y_pred = classifier_model.predict(X_test)

#Step 6: Display the performance on the testing set
print ('Accuracy Score :',accuracy_score(y_test, y_pred))
print (metrics.classification_report(y_test, y_pred))
print ('roc_auc_score: ', roc_auc_score(y_test, y_pred))

#Step 7: Present the metrics
visualisation = ClassificationReport(classifier_model, support=True)
visualisation.fit(X_train, y_train)
visualisation.score(X_test, y_test)          # Assess the model on the test data
visualisation.show()                         # Show the figure

#Step 8: Plot the confusion matrix
matrix = confusion_matrix(y_test, y_pred)
labels = ['True Negative','False Positive','False Negative','True Positive']
label_counts = ['{0:0.0f}'.format(value) for value in
                matrix.flatten()]
label_percent = ['{0:.2%}'.format(value) for value in
                 matrix.flatten()/np.sum(matrix)]
labels = [f'{v1}\n{v2}\n{v3}' for v1, v2, v3 in
          zip(labels, label_counts, label_percent)]
labels = np.asarray(labels).reshape(2,2)
sns.heatmap(matrix, annot=labels, fmt='', cmap='Blues')
```

Fig. 4.15 Simple machine learning codes to perform decision tree classification (flawed approach – but do see the steps and codes, as those are correct)

```
Accuracy Score : 0.8074534161490683
              precision    recall  f1-score   support

           0       0.86      0.86      0.86       219
           1       0.70      0.70      0.70       103

    accuracy                           0.81       322
   macro avg       0.78      0.78      0.78       322
weighted avg       0.81      0.81      0.81       322

roc_auc_score:  0.7787383073990335
```

Fig. 4.16 The confusion matrix and ROC AUC score (outputs to evaluate an ML model performance)

predictions with the actual truth (y_test) and report the error rates in step 6 (see below).

Step 6: We will evaluate how well the model learnt using accuracy score, confusion matrix (using classification_report) and roc_auc_score as shown in Fig. 4.16.

We found an accuracy score of 0.81. This means that out of all the predictions made by the decision tree, 81% of the results were correct.

The ROC AUC score is 0.78. But what is it?

Short answer: It's like saying the probability that the classifier will be able to differentiate between the positive and negative values is 78/100. The higher the value, the closer it is to 1 and the better it is. The result means that the current predictor model is better than a random coin toss (which also has a chance of being right 50% of the time).

Now the long answer: The full form of AUC is *area under curve*. For ROC, it is *receiver operating characteristic*. Combined, AUC ROC refers to the area under the curve of the ROC curve. Independently, the ROC curve informs us how well the classification model can distinguish between two classes. Our example in this lesson had been if a person has demonstrated a target behaviour or not. Good classification models can accurately distinguish between the two (i.e. YES and NO) better than a bad model. The score representing the AUC is the area under the ROC curve that gives us a fair idea of how well the model performs. If we have got an AUC ROC score of 0.8 (for example), that would have meant the classifier could distinguish the positive and the negative values fairly well. We can use AUC ROC metrics as a performance indicator between two or more classifier models. We aim to compare the performances and pick the best (as we did with the F1 score in the previous section).

Which metric is better: AUC ROC or accuracy?

1. If you have a balanced classification problem, approximately 50% of a binary target variable is labelled as class A (e.g. depressed) and the other 50% as class B (e.g. not depressed), then accuracy could be the best measure.
2. For an imbalanced classification problem, where the distribution of data in the target variable is not exactly 50-50 or even 60-40 (as above), and rather like 80% of the data values belong to class A and only 20% to class B, then AUC is a better metric than accuracy, as in this case, accuracy is influenced by the majority class.

Step 7 (optional): You can visualise the classification report indicating the model's performance. As shown in the plot above, the precision, recall and f1 while predicting 1 (YES to suicide attempt) are zeros.

Bonus

It is important to ensure that your target variable is balanced. Otherwise, the model will perform poorly. This means that the number of 1 and the number of 0 should be near about 50-50. It should not be 90-10 or even 80-20. This reflects in the support score. The more balanced the target variable is, the better the model will be able to learn. This is somewhat obvious because imagine that I tell you to identify two odours: scent A and scent B. I presented you scent A nine times while I presented scent B only once. Imagine that the order was A, A, B, A, A, A, A, A, A, A. Will you be able to recognise scent B accurately in the future? Probably not. So, I must ensure that you smell scent A five times and scent B five times. In real-world data situations when such a perfect division of 50/50 is not possible, we must at least try to get as close to that (e.g. 60/40).

Classification Report

Now, look at the plot in Fig. 4.17. We are estimating the precision, recall, F1 and support for each class: 0 and 1. If you recall, 1 indicated YES (target behaviour occurred), and 0 indicated NO (target behaviour did not occur).

What Does Each of These Means?

- **Precision** (classifiers' exactness: for all predictions classified as positive, what percentage was correct?). We can calculate precision as = True Positives/(True Positives + False Positives). We got the values for true positives and false positives in the confusion matrix.
- **Recall** (classifiers' completeness or **sensitivity**): It answers for all actually positive predictions; what percentage was classified correctly? We can estimate Recall by = True Positives/(True Positives + False Negatives). Just enter the values from the confusion matrix above in the formula.
- **F1 score** (the best score is 1.0, and the worst is 0.0). The weighted harmonic mean of precision and recall (above 2). We used the weighted average of F_1 to compare classifier models (instead of accuracy). You can do that, too, as a rule of thumb. Choose the model which has the highest F1 score for the dataset. It's easy to calculate the F1 score. From the above two, we already got precision and recall. Now F1 = (2 * Precision * Recall)/(Precision + Recall).
- **Support** (the number of actual class occurrences in the testing dataset). The best possible support figures for each class (which are 1 and 0) would be an equal number each. In simpler words, the machine learning algorithm loves it when out of two target classes, each one is divided 50–50%. This might not be possi-

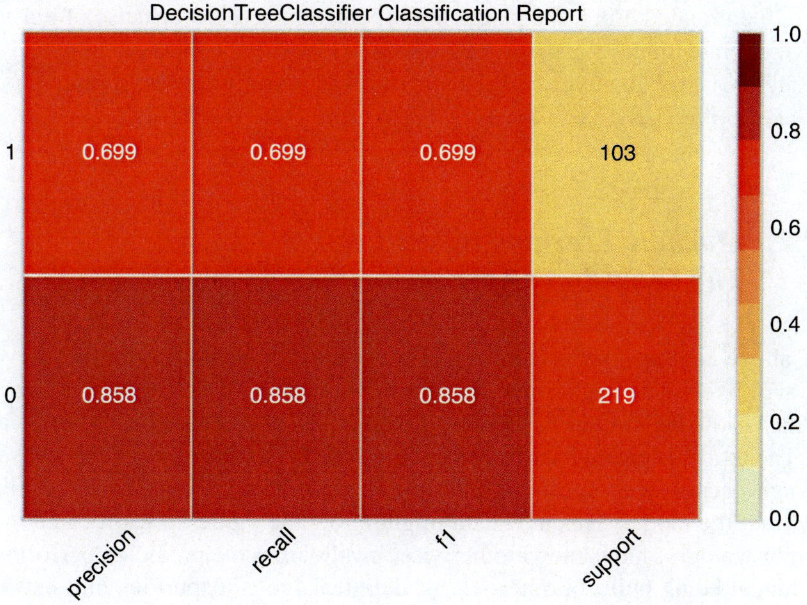

Fig. 4.17 Classification report

ble in the real world. But closer the figure is to 50%, the better it is. It would be reasonably stupid to attempt to predict YES and NO responses based on a dataset where 98% of people did not show the target behaviour (NO: 0) and only 2% of people demonstrated the target behaviour (YES: 1). The opposite is true as well. Comparing YES responses from 98% of the participants with 2% of NO responses is equally ridiculous. So, the support parameters show us if we have got an unbalanced dataset (target).

Step 8 (optional): You can plot the confusion matrix to calculate the sensitivity and specificity of each model. In this context, sensitivity means the proportion of people who were predicted to attempt suicide out of those who actually have attempted suicide. Specificity means the proportion of people who were predicted to NOT attempt suicide out of those who do not actually attempt suicide.

- Sensitivity = True Positive/(True Positive + False Negative)
- Specificity = True Negative/(False Positive + True Negative)

Accordingly, in the case of the current findings, it seems like our model's sensitivity and specificity (in terms of percentages) are 69.90% and 85.84%. You can build multiple models, and by comparing the sensitivity and specificity scores, you can decide which model is kept for further investigation (you can delete or remove others).

Confusion Matrix

Bonus
As mentioned before, the resultant outputs are not good because of our data. I have intentionally used toy data that is small enough to demonstrate the nature of problems you will end up with if you do not use a large enough dataset (Fig. 4.18).

4.2.2 Machine Learning (Classifier): The Leak-Proof Approach

The above approach is perfect, except for one element. If you run the above code, you will likely get caught up in a serious data leakage problem. As we saw earlier, the data is divided into two subsets in supervised machine learning: train set and test set. The model is created using the training set and evaluated using the test set. Data leakage occurs when information from outside the training dataset is utilised to construct the model. This new data might allow the model to learn or know something it wouldn't have known otherwise, invalidating the predicted performance of the model being built. In other words, data leakage is a spurious link between the independent and target variables that develops due to the data collection, sampling

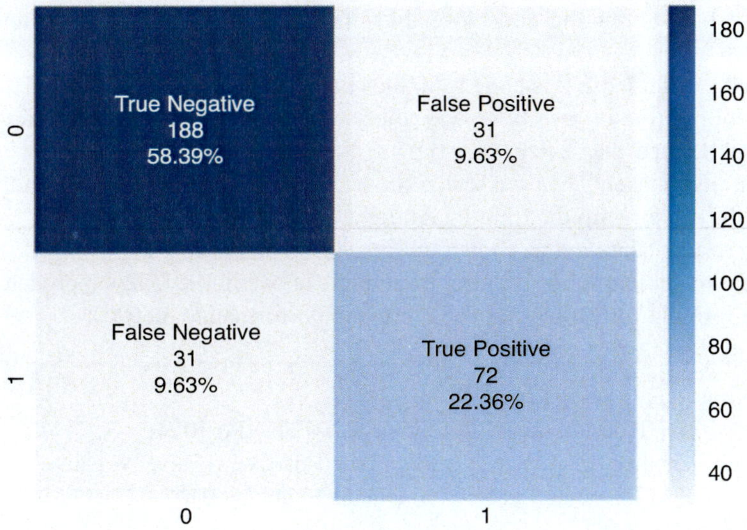

Fig. 4.18 The confusion matrix

or pre-processing technique. As a result, data leakage might lead to over-optimistic performance claims. To prevent this from happening, you must take specific measures such as the following:

- One of the most common ways in which data leakage happens is when the missing values are imputed for both the training and test data together. In related terms, another way to leak data is to reuse an imputed dataset incorrectly. To prevent both, remove all the missing values from the dataset (hence no need for imputation). Or perform imputation separately for the training dataset and test dataset. The above code did not have this problem because we removed all the rows with missing values. However, suppose you don't have enough data and try to impute the missing values. In that case, you must perform it separately for training and test dataset.
- Likewise, when performing MinMax scaling or standardisation, make sure to perform them on the training dataset and test the dataset independently. This is what we did not do above in the "stupid approach", and many blogs and free video tutorials on YouTube make this same mistake. In the codes below, you will see how easily you can solve this problem with just three lines of code.
- The other type of error made causing data leakage is by using proxy variables. The usual practice is that the target variable is assigned to the model. Then all the relevant variables in the dataset apart from that target variable are selected as independent variables or predictors (and assigned to the models). Now, if the target variable is "sleep disturbance", and if one of the predictors is "difficulties in falling asleep", then that will be called a "proxy variable", and it can cause data leakage leading to over-optimistic performance claims. Our demonstration carefully noted that none of the independent variables or predictors was a proxy for the target variable.

So, what should the code look like if we consider the data leakage aspect? (Fig. 4.19)

What's different in this code from the "naïve" approach before is that the marked red portion in the codes above was after we split the dataset into training and test data, we defined the scaler (i.e. `MinMaxScaler()`). Subsequently, we fit it into the training dataset. Then we scaled the training and test datasets separately before fitting the model with the training dataset.

The results came out as shown in Figs. 4.20, 4.21 and 4.22.

You already learnt how to infer these plots above in the leaky approach. But here are the formula for estimating the following performance metrics:

- Accuracy = (True Positive + True Negative)/(True Positive + True Negative + False Positive + False Negative)
- Precision = True Positive/(True Positive + False Positive)
- Recall (sensitivity) = True Positive/(True Positive + False Negative)
- F1 Score = (2* recall * precision)/(recall + precision)

If you don't remember, then go back to the previous pages where we discussed the "Leaky" approach and check how you can estimate the sensitivity and specificity of the models. If you are not happy with the model's performance, try another model or change the model's hyperparameters (as you will see in the upcoming sections). Another aspect to check is if you processed the dataset appropriately.

When you are happy with your machine learning model's performance, you can use it to implement it in a web app or software. Every time you enter an input (independent/predictor variables), you get output about the target variable. You can use

```
In [42]: #Assign the target (dependent) variable and the independent variables
feature_names = ["Binary_Gender_Male", "Age", "Not being able to stop or control worrying", "Little interest or pleasure in doin
X = data[feature_names]     #predictors
y = data["Binary_Feeling nervous, anxious or on edge__Low"] #predicted
X_train, X_test, y_train, y_test = train_test_split(X, y, test_size=0.30, random_state=100)

#re-scale the variables
Variable_Scaler = MinMaxScaler()
Variable_Scaler.fit(X_train)
X_train = Variable_Scaler.transform(X_train)
X_test = Variable_Scaler.transform(X_test)

classifier_model = DecisionTreeClassifier()
classifier_model=classifier_model.fit(X_train, y_train)
y_pred = classifier_model.predict(X_test)

#Display the performance on the testing set
print ('Accuracy Score :',accuracy_score(y_test, y_pred))
print (metrics.classification_report(y_test, y_pred))
print ('roc_auc_score: ', roc_auc_score(y_test, y_pred))

#Present the metrics
visualisation = ClassificationReport(classifier_model, support=True)
visualisation.fit(X_train, y_train)
visualisation.score(X_test, y_test)      # Assess the model on the test data
visualisation.show()                     # Show the figure

#Plot the confusion matrix
matrix = confusion_matrix(y_test, y_pred)
labels = ['True Negative','False Positive','False Negative','True Positive']
label_counts = ['{0:0.0f}'.format(value) for value in
                matrix.flatten()]
label_percent = ['{0:.2%}'.format(value) for value in
                 matrix.flatten()/np.sum(matrix)]
labels = [f'{v1}\n{v2}\n{v3}' for v1, v2, v3 in
          zip(labels, label_counts, label_percent)]
labels = np.asarray(labels).reshape(2,2)
sns.heatmap(matrix, annot=labels, fmt='', cmap='Blues')
```

Fig. 4.19 The flawless approach to running a machine learning model

```
Accuracy Score : 0.8012422360248447
              precision    recall  f1-score   support

           0       0.85      0.85      0.85       219
           1       0.69      0.69      0.69       103

    accuracy                           0.80       322
   macro avg       0.77      0.77      0.77       322
weighted avg       0.80      0.80      0.80       322

roc_auc_score:  0.7716008334441637
```

Fig. 4.20 The output of the machine learning model, part 1

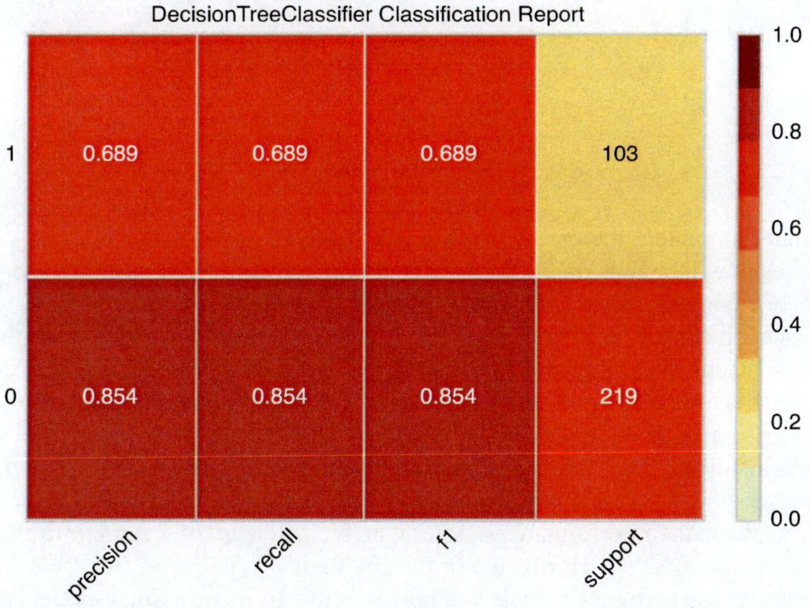

Fig. 4.21 Output of the machine learning model, part 2

such a tool in your clinical practice. As you might imagine, such a machine learning application interests a software developer more than a psychologist interested in using machine learning for research. So, I won't go in depth about building a mobile phone or website app that uses this machine learning model (that would cater to a different audience). That would be outside the scope of this book. We are trying to learn how to use machine learning to analyse your data in academic research and learn more about the phenomenon. Therefore, we will do something else, something more interesting. Instead of predicting the target variable, we will learn to reverse engineer this model and see which predictor variables contribute to the target prediction. That will indicate which variables are important while studying the target phenomena or behaviour. Therefore, I will focus on how you can use machine

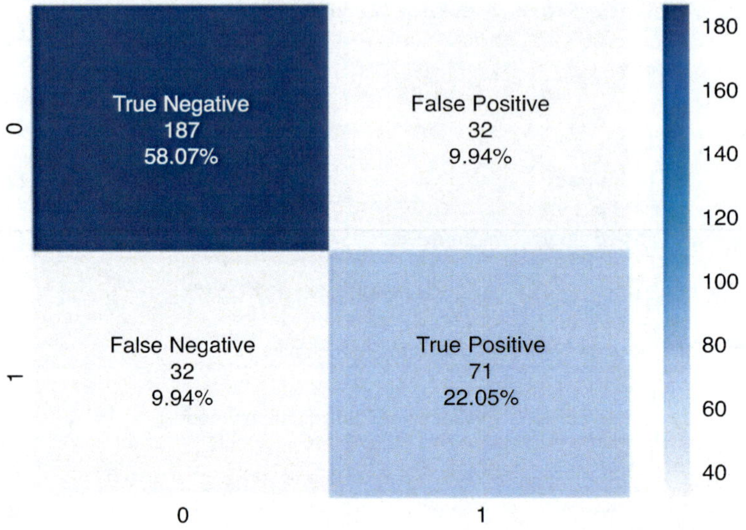

Fig. 4.22 Output of the machine learning model (confusion matrix), part 3

learning to explore the extent to which each predictor predicts the target variable. The result will be that you will learn about the predictors of your target phenomena or behaviour and contribute towards advancing its literature.

That's it for this section. I hope the above example introduced you to the basic idea of machine learning and its use in the industry. The subsequent discussion will be done keeping in mind that this book is intended for researchers who wish to implement machine learning to analyse their data and not build an app. So, we will learn this during the lesson on feature importance. But before that, we will learn more about how to run multiple machine learning models within one code script and then evaluate their performances. After that, we will learn how to optimise the best-performing model's performance further by tuning hyperparameters. We will also see how to run regressor models and how it differs from a classifier model. Then we will finally get down to learning how to find feature importance.

As you can see from the comparison table in Table 4.3, the difference in the performance of both models is not substantial. However, any instance of data leakage dilutes the significance of your findings, and you wouldn't want that. Right?

4.2.3 Confidence Interval

It is highly recommended that you include significance testing or uncertainty quantification to compare the models' predictive performance. A review found that 9 of the 12 recent studies published in top political science journals for which complete code and data were available included no significance tests or uncertainty quantification for classifier performance comparison (Kapoor & Narayanan, 2021).

Table 4.3 Difference between a leaky approach and a non-leaky approach

<table>
<tr>
<th>Leaky approach
(Rescale the whole dataset and THEN build a machine learning model)</th>
<th>Recommended approach
(Rescaling the training set and test set separately and THEN building the machine learning model)</th>
</tr>
<tr>
<td><pre>
In [565]: Variable_Scaler = MinMaxScaler()
          scaled_data = pd.DataFrame(Variable_Scaler.fit_transform(data))

          #Print the normalised data
          print (scaled_data.head())

          #Exporting the normalised data to your computer
          scaled_data.to_csv('Scaled_Dataset_Tutorial.csv')

          0   1   2   3   4   5   6
0 0.310345 1.000000 1.000000 1.000000 0.333333 0.0 1.0
1 0.379310 1.000000 0.666667 0.666667 0.000000 0.0 1.0
2 0.206897 1.000000 0.666667 0.666667 1.000000 0.0 1.0
3 0.482759 0.666667 1.000000 1.000000 0.666667 0.0 1.0
4 0.362069 1.000000 1.000000 1.000000 1.000000 0.0 1.0
</pre></td>
<td>We will do MinMaxScaler() after separating the training and test datasets, in this case, as shown below</td>
</tr>
<tr>
<td><pre>
#Assign variables as predictors and target
feature_names = ["Binary_Gender_Male", "Age", "Not being able to stop or control worrying",
                 "Little interest or pleasure in doing things", "Feeling down, depressed, or hopeless"]
X = data[feature_names]   #predictors
y = data["Binary_Feeling nervous, anxious or on edge__Low"] #predicted

#Divide the dataset into two parts: training set and test set.
X_train, X_test, y_train, y_test = train_test_split(X, y, test_size=0.30, random_state=100)

#Build the model
classifier_model = DecisionTreeClassifier()
classifier_model=classifier_model.fit(X_train, y_train)
y_pred = classifier_model.predict(X_test)

#Display the performance on the testing set
print ('Accuracy Score :',accuracy_score(y_test, y_pred))
print (metrics.classification_report(y_test, y_pred))
print ('roc_auc_score: ', roc_auc_score(y_test, y_pred))

#Present the metrics
visualisation = ClassificationReport(classifier_model, support=True)
visualisation.fit(X_train, y_train)
visualisation.score(X_test, y_test)   # Assess the model on the test data
visualisation.show()   # Show the figure

#Plot the confusion matrix
matrix = confusion_matrix(y_test, y_pred)
labels = ['True Negative','False Positive','False Negative','True Positive']
label_counts = ['{0:0.0f}'.format(value) for value in
                matrix.flatten()]
label_percent = ['{0:.2%}'.format(value) for value in
                 matrix.flatten()/np.sum(matrix)]
labels = [f'{v1}\n{v2}\n{v3}' for v1, v2, v3 in
          zip(labels, label_counts, label_percent)]
labels = np.asarray(labels).reshape(2,2)
sns.heatmap(matrix, annot=labels, fmt='', cmap='Blues')
</pre></td>
<td><pre>
#Assign the target (dependent) variable and the independent variables
feature_names = ["Binary_Gender_Male", "Age", "Not being able to stop or control worrying", "Little interest or pleasure in doing
X = data[feature_names]   #predictors
y = data["Binary_Feeling nervous, anxious or on edge__Low"] #predicted
X_train, X_test, y_train, y_test = train_test_split(X, y, test_size=0.30, random_state=100)

#re-scale the variables
Variable_Scaler = MinMaxScaler()
Variable_Scaler.fit(X_train)
X_train = Variable_Scaler.transform(X_train)
X_test = Variable_Scaler.transform(X_test)

classifier_model = DecisionTreeClassifier()
classifier_model=classifier_model.fit(X_train, y_train)
y_pred = classifier_model.predict(X_test)

#Display the performance on the testing set
print ('Accuracy Score :',accuracy_score(y_test, y_pred))
print (metrics.classification_report(y_test, y_pred))
print ('roc_auc_score: ', roc_auc_score(y_test, y_pred))

#Present the metrics
visualisation = ClassificationReport(classifier_model, support=True)
visualisation.fit(X_train, y_train)
visualisation.score(X_test, y_test)   # Assess the model on the test data
visualisation.show()   # Show the figure

#Plot the confusion matrix
matrix = confusion_matrix(y_test, y_pred)
labels = ['True Negative','False Positive','False Negative','True Positive']
label_counts = ['{0:0.0f}'.format(value) for value in
                matrix.flatten()]
label_percent = ['{0:.2%}'.format(value) for value in
                 matrix.flatten()/np.sum(matrix)]
labels = [f'{v1}\n{v2}\n{v3}' for v1, v2, v3 in
          zip(labels, label_counts, label_percent)]
labels = np.asarray(labels).reshape(2,2)
sns.heatmap(matrix, annot=labels, fmt='', cmap='Blues')
</pre></td>
</tr>
<tr>
<td>Accuracy Score: 0.81
ROC AUC score: 0.78</td>
<td>Accuracy Score: 0.80
ROC AUC score: 0.77</td>
</tr>
</table>

```
In [23]: import pandas as pd
         data = pd.read_csv("C:\\Users\\ghosh\\Downloads\\Pre-processed dataset.csv", header=0, index_col=0)

In [24]: from sklearn.model_selection import train_test_split
         #Assign the target (dependent) variable and the independent variables
         feature_names = ["Binary_Gender_Male", "Age", "Not being able to stop or control worrying", "Little interest or pleasure in doing
         X = data[feature_names]     #predictors
         y = data["Binary_Feeling nervous, anxious or on edge__Low"] #predicted
         X_train, X_test, y_train, y_test = train_test_split(X, y, test_size=0.30, random_state=100)

In [25]: from sklearn.utils import resample
         from matplotlib import pyplot
         import numpy
         from sklearn.tree import DecisionTreeClassifier
         from sklearn.metrics import accuracy_score

         number_of_iterations = 100
         sample_size = int(len(data) * 0.65)
         creating_list = list()

         for c in range(number_of_iterations):
             # prepare train and test sets
             New_train = resample(X_train, n_samples=sample_size)
             train = resample(y_train, n_samples=sample_size)
             y_test_new = resample(y_test, n_samples=sample_size)
             X_test_new = resample(X_test, n_samples=sample_size)
             test = numpy.array([x for x in X_train if x not in train])
             # fit model
             model = DecisionTreeClassifier().fit(New_train, train)
             # evaluate model
             predictions = model.predict(X_test_new)
             score = accuracy_score(y_test_new, predictions)
             print(score)
             creating_list.append(score)
         #plotting the scores
         pyplot.hist(creating_list)
         pyplot.show()
         #estimating 95% confidence interval of the model
         alpha = 0.95
         p = ((1.0-alpha)/2.0) * 100
         lower = max(0.0, numpy.percentile(creating_list, p))
         p = (alpha+((1.0-alpha)/2.0)) * 100
         upper = min(1.0, numpy.percentile(creating_list, p))
         print('95 per cent confidence interval %.1f%% and %.1f%%' % (lower*100, upper*100))
```

Fig. 4.23 Code to estimate the confidence interval of the model

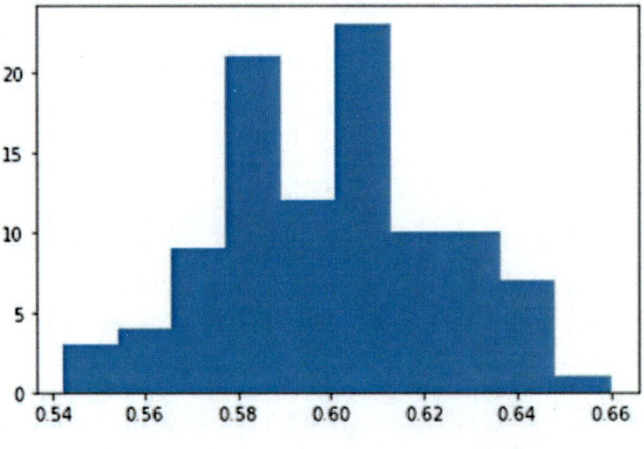


```
95 per cent confidence interval 55.4% and 64.4%
```

Fig. 4.24 The confidence interval of the model

However, it's easy to estimate the confidence interval of the model and can use the code in Fig. 4.23. You will get an output as shown in Fig. 4.24.

Note that even when we had different independent variables or predictors on different scales, we did not use scaled data in this case as we are using the decision tree model. However, suppose you wish to make non-tree-based approaches such as logistic regression. In that case, it is better to rescale or standardise/normalise the data for the reasons mentioned in this chapter.

4.2.4 Choosing the Best Model for Classification

In the previous section, you got an overview of a machine learning prediction script. We loaded the data, split it into training and testing sets, fitted a classifier model to the training data, made predictions based on this data and tested the predictions on the test data. But train/test split does have its dangers – because what if the split we made isn't random and one subset of our data has only people from certain demography (e.g. age, gender and income level) – biased data. Such attributes will result in overfitting, which we are supposed to be avoiding. This is where cross-validation comes in. In this section, I will demonstrate the cross-validation method.

Furthermore, the current lesson also attempts to help you answer the question: many models exist. We picked the decision tree above. But there are others as well. So, which one to pick?

The answer is to try as many of them as possible to check which one performs the best. The best part is we can do this all in a single code script, and no need to run the algorithms one by one on different task screens. See the screen below for a demonstration of the classification task using some of the most commonly used classifier algorithms (there are more):

1. Logistic regression
2. Decision tree classifier
3. Random forest classifier
4. K-neighbours classifier
5. Linear discriminant analysis
6. Gaussian naive Bayes
7. Multi-layer perceptron classifier

Check the Code Below for How to Do That (Fig. 4.25):

Bonus

- Decision tree, random forest, K-neighbours and multi-layer perceptron algorithms can be used both for classification (i.e. predict categorical classes) and regression (i.e. predict continuous numbers) tasks.
- Logistic regression, linear discriminant analysis and Gaussian naive Bayes are exclusively for classification. They cannot be used for regression.
- Tree-based models (such as decision tree and random forest) and artificial neural network (such as multi-layer perceptron) do not require feature rescaling or standardisation of the dataset. They are also some of my favourites because of their superior performance in several instances.

(continued)

- Suppose your model prediction accuracy comes at 1.0. This means the model is accurate 100% of the time. But, before you get all excited about it, no prediction is 100% accurate. If it shows 1.0, then chances are we have done something wrong. Maybe, we need to tune the hyperparameters. Maybe, we need to go back to the pre-processing stage and check for other strategies. Maybe, there are some other issues such as overfitting. But in any case, 100% accuracy is flawed – both in life and machine learning literature.
- Multi-layer perceptron classifier is an example of a deep learning algorithm. **How is it connected to machine learning?** Deep learning is a subset of machine learning.
  - You must have heard about the term neural network in connection to the human brain. Deep learning is inspired by the working of our brain. It utilises artificial neural network to solve complex machine learning issues.
  - Artificial neural network works better when we perform Z-score normalisation (using StandardScaler() utility class, Page: 73) because normalising the data usually speeds up the learning and leads to faster convergence. But remember, our current data were rescaled using MinMaxScaler. So, in your real-world application, you might come across such situations. So, ensure that you use the StandardScaler utility on the raw data. The output of traditional machine learning algorithms is usually numeric. But the output of deep learning can be anything. For example, it can generate music and images and can write paragraphs of texts on its own.

Heard about deepfake and the controversies surrounding it? If not, I recommend you search it online right now. You will be fascinated with technology. Other examples of deep learning applications are self-driving cars, social robots and next-generation software. But we, psychological scientists, are not building our own self-driving car. Instead, we will utilise deep learning for data analysis. But note, although you might have been impressed by deep learning, that does not imply that it will perform better than other machine learning algorithms. So, your preference for the model should be based on the performance and not the hype.

```
In [13]: #Step 1: Import libraries, packages and modules
import pandas as pd
import matplotlib.pyplot as plt
from sklearn.model_selection import train_test_split
from sklearn import model_selection
from sklearn.preprocessing import MinMaxScaler
from sklearn.linear_model import LogisticRegression
from sklearn.tree import DecisionTreeClassifier
from sklearn.ensemble import RandomForestClassifier
from sklearn.naive_bayes import GaussianNB
from sklearn.neighbors import KNeighborsClassifier
from sklearn.discriminant_analysis import LinearDiscriminantAnalysis
from sklearn.neural_network import MLPClassifier
from sklearn.pipeline import Pipeline

#Step 2: Load the dataset
data = pd.read_csv("C:\\Users\\ghosh\\Downloads\\Pre-processed dataset.csv", header=0, index_col=0)

#Step 3: Assign the target (dependent) variable and the predictor variables
variable_names = ["Binary_Gender_Male", "Age", "Not being able to stop or control worrying", "Little interest or pleasure in doi
X = data[variable_names]     #predictors
y = data["Binary_Feeling nervous, anxious or on edge__Low"] #predicted
X_train, X_test, y_train, y_test = train_test_split(X, y, test_size=0.30, random_state=0)

#Step 5: Build the prediction models
models = []
models.append(('Logistic Regression', LogisticRegression()))
models.append(('DecisionTreeClassifier', DecisionTreeClassifier()))
models.append(('RandomForestClassifier', RandomForestClassifier()))
models.append(('Gaussian Naive Bayes', GaussianNB()))
models.append(('K Neighbors Classifier', KNeighborsClassifier()))
models.append(('Linear Discriminant Analysis', LinearDiscriminantAnalysis()))
models.append(('Multi-layer Perceptron classifier', MLPClassifier(max_iter=1000)))

#Step 6: Evaluate all the models to compare and choose the best
targets = []
algorithm_names = []

for ML_models, model in models:
    kfold = model_selection.KFold(n_splits=10, shuffle=True, random_state=0)
    Variable_Scaler = MinMaxScaler()
    pipeline = Pipeline(steps=[('VariableScaler', Variable_Scaler), ('Model', model)])
    cv_results = model_selection.cross_val_score(model, X, y, cv=kfold, scoring='f1_weighted') #Other scoring metrics available
    targets.append(cv_results)
    algorithm_names.append(ML_models)
    output = "%s: %f (%f)" % (ML_models, cv_results.mean(), cv_results.std())
    print(output)

#step 7: Box and whisker plot to visualise the model performance
Box_and_whisker_plot = plt.figure()
Box_and_whisker_plot.suptitle('Comparing the ML models in terms of accuracy')
ax = Box_and_whisker_plot.add_subplot()
plt.boxplot(targets)
ax.set_xticklabels(algorithm_names, rotation=90)
plt.show()
```

Fig. 4.25 Codes to run multiple classifier algorithms to compare their performances and pick the most accurate model for further work

Evaluating Model Performance

From Fig. 4.26, it is clear that the logistic regression is the most accurate model compared to others on this dataset (i.e. 0.88). We chose "F1" as the scoring metric to evaluate the performance of models. But there are several others. Search the following link for the list (https://scikit-learn.org/stable/modules/model_evaluation.html#scoring-parameter).

Now that we know the logistic Regression has the highest weighted average of the precision and recall (F1 score), we can use the same code from the non-leaky approach (page: 114), but this time with Logistic Regression (instead of the decision tree). Suppose you have a strong reason to choose one algorithm over others. In that case, you can skip the comparison step (as above) and directly go for the codes given under a non-leaky approach.

Fig. 4.26 Comparative performance of different classifier models

```
Logistic Regression: 0.881778 (0.023105)
DecisionTreeClassifier: 0.798221 (0.034458)
RandomForestClassifier: 0.836801 (0.031569)
Gaussian Naive Bayes: 0.852748 (0.027943)
K Neighbors Classifier: 0.846345 (0.031112)
Linear Discriminant Analysis: 0.877081 (0.023209)
Multi-layer Perceptron classifier: 0.882200 (0.025082)
```

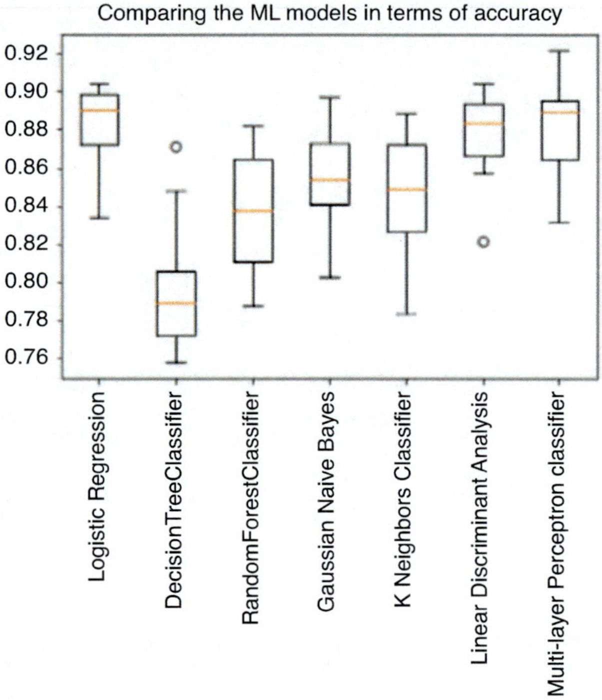


```
In [126]: import warnings
          warnings.filterwarnings("ignore")
```

Fig. 4.27 Codes to ignore the warnings

Bonus

There might be times when you can get “warnings” in Python. Warnings are not errors. Usually, Python throws up warnings for the developers notifying the user of certain aspects of the code. Examples might include how the upcoming updates will change certain default values if something we have used works. Still, there is an updated version available and so forth. Other times, warnings can hint where your code might need some modification for better results. If you want to ignore it, simply enter the code as shown in Fig. 4.27 in your script.

4.2.5 Optimising the Predictive Accuracies of the Model with Hyperparameter Tuning

In the previous lesson (Page: 121), we learnt how to choose the best classifier model using cross-validation. You can do exactly the same for regressors (as seen in the next lesson). So, whether it is classifiers or regressors, cross-validation helps to evaluate the different models and pick the one that is the best performing. But is there a way to improve the model performances? We will learn that in this section.

You must have noticed that each model has numerous settings in them. These settings vary for different models. But combined, they are called hyperparameters. The process of choosing the optimal set of values is called hyperparameter tuning (or optimisation). For the random forest model, examples of such hyperparameters are the number of trees in the forest of the model, the maximum depth of each tree, the minimum number of samples required to split an internal leaf node and so forth.

If we do not specify them, they are set to default values. But if we set them optimally, chances are it will perform even better. Let me show you what I mean. If you remember the previous section, we specified the K-neighbours classifier and two parentheses. We could have entered the parameters inside it, but we left it blank. So, leaving it blank means leaving it to default values.

```
models.append(('K Neighbors Classifier', KNeighborsClassifier()))
```

Likewise, go back to the previous lesson where we chose the best-performing classifier (or just see the codes in Fig. 4.28). You will notice that most of the classifiers are left to default settings.

We found logistic regression to be the most accurate but just for variation, let us assume that we found the K-neighbours classifier as the best possible model (using cross-validation). But we can further improve the performance of the K-neighbours classifier if we go from default settings to optimum setting values. But how do we know those best possible combinations? There are large but finite possibilities. So, we opt for different strategies to automate the process of finding the best possible combination of hyperparameters. Note, that hyperparameters are set manually while parameters are learnt automatically. Therefore, both the terms are not synonymous.

In this section, we will learn about two of the most popularly used methods:

```
#Step 5: Build the prediction models
models = []
models.append(('Logistic Regression', LogisticRegression()))
models.append(('DecisionTreeClassifier', DecisionTreeClassifier()))
models.append(('RandomForestClassifier', RandomForestClassifier()))
models.append(('Gaussian Naive Bayes', GaussianNB()))
models.append(('K Neighbors Classifier', KNeighborsClassifier()))
models.append(('Linear Discriminant Analysis', LinearDiscriminantAnalysis()))
models.append(('Multi-layer Perceptron classifier', MLPClassifier(max_iter=1000)))
```

Fig. 4.28 Codes to build the machine learning classifier models

Strategy 1: Exhaustive Grid Search

```
In [131]: #Step 1: Import libraries, packages and modules
from sklearn.neighbors import KNeighborsClassifier
from sklearn.model_selection import train_test_split
from sklearn.model_selection import GridSearchCV
import pandas as pd

#Step 2: Load the dataset
data = pd.read_csv("C:\\Users\\ghosh\\Downloads\\Pre-processed dataset.csv", header=0, index_col=0)
data.head()

#Step 3: Assign the target (dependent) variable and the predictor variables
variable_names = ["Binary_Gender_Male", "Age", "Not being able to stop or control worrying", "Little interest or pleasure in doi
X = data[variable_names]     #predictors
y = data["Binary_Feeling nervous, anxious or on edge__Low"] #predicted

#Step 4: Splitting the data into test and train dataset
X_train, X_test, y_train, y_test = train_test_split(X, y, test_size=0.30)

#Step 5: Re-scale the variables
Variable_Scaler = MinMaxScaler()
Variable_Scaler.fit(X_train)
X_train = Variable_Scaler.transform(X_train)
X_test = Variable_Scaler.transform(X_test)

#Step 6: Build the prediction classifier model
KNeighbors_Classifier_Model = KNeighborsClassifier()
KNeighbors_Classifier_Model.fit(Xtrain, ytrain)

#Step 7: Specifying the search space
Grid_search = {'n_neighbors': [1,3, 5, 7,  9],
               'weights': ['uniform', 'distance'],
               'algorithm': ['auto', 'ball_tree', 'kd_tree', 'brute'],
               'metric': ['euclidean', 'manhattan']}

#Step 8: Perform random search based on the specified search space
KNeighbors_Classifier_Model = GridSearchCV(estimator = KNeighbors_Classifier_Model,
                                  param_grid=Grid_search, verbose = 1,
                                  cv = 3)

#Step 9: Fit the random search model
KNeighbors_Classifier_Model.fit(Xtrain, ytrain)

#Step 10: Print the best options
print (KNeighbors_Classifier_Model.best_score_)
print (KNeighbors_Classifier_Model.best_estimator_)
print (KNeighbors_Classifier_Model.best_params_)

Fitting 3 folds for each of 80 candidates, totalling 240 fits
0.8442231075697211
KNeighborsClassifier(algorithm='brute', metric='manhattan', n_neighbors=7)
{'algorithm': 'brute', 'metric': 'manhattan', 'n_neighbors': 7, 'weights': 'uniform'}
```

Fig. 4.29 Codes to perform grid search

1. **Grid search**
2. **Random search**

In both the search methods, we need to define a search space (volume to be searched) in advance, and the search functions will perform within it. See the codes and outputs below for each. That will make it clearer.

Grid Search

In Grid search, the machine does an *exhaustive search* of the hyperparameters and finds the best ones. To do this, we first need to specify the parameters we want to search. After that, `GridSearchCV` will perform all the required model fits and throw us the result (Fig. 4.29).

Random Search

Random search does not do an exhaustive search of all combinations (like GridSearch does). Instead, it randomly selects a fixed number of parameter settings from the specified distributions. Unlike the GridSearch method, not all parameter values are tried out here. You will notice that there is a **n_iter** hyperparameter for a random search. If you set it to 100, then it would try random combinations 100

Strategy 2: Randomised Parameter Optimisation

```
In [132]: #Step 1: Import Libraries
from sklearn.neighbors import KNeighborsClassifier
from sklearn.model_selection import train_test_split
from sklearn.model_selection import RandomizedSearchCV
import pandas as pd

#Step 2: Load the dataset
data = pd.read_csv("C:\\Users\\ghosh\\Downloads\\Pre-processed dataset.csv", header=0, index_col=0)
data.head()

#Step 3: Assign the target (dependent) variable and the predictor variables
variable_names = ["Binary_Gender_Male", "Age", "Not being able to stop or control worrying", "Little interest or pleasure in doi
X = data[variable_names]     #predictors
y = data["Binary_Feeling nervous, anxious or on edge__Low"] #predicted

#Step 4: Splitting the data into test and train dataset
Xtrain, Xtest, ytrain, ytest=train_test_split(X, y, test_size=0.30)

#Step 5: Re-scale the variables
Variable_Scaler = MinMaxScaler()
Variable_Scaler.fit(X_train)
X_train = Variable_Scaler.transform(X_train)
X_test = Variable_Scaler.transform(X_test)

#Step 6: Build the prediction classifier model
KNeighbors_Classifier_Model = KNeighborsClassifier()
KNeighbors_Classifier_Model.fit(Xtrain, ytrain)

#Step 7: Specifying the search space
random_search  = {'n_neighbors': [1,3, 5, 7,  9],
               'weights': ['uniform', 'distance'],
               'algorithm': ['auto', 'ball_tree', 'kd_tree', 'brute'],
               'metric': ['euclidean', 'manhattan']}


#Step 8: Perform random search based on the specified search space
KNeighbors_Classifier_Model = RandomizedSearchCV(estimator = KNeighbors_Classifier_Model,
                                param_distributions = random_search, n_iter = 80, verbose = 1,
                                cv = 3)

#Step 9: Fit the random search model
KNeighbors_Classifier_Model.fit(Xtrain, ytrain)

#Step 10: Print the best options
print (KNeighbors_Classifier_Model.best_score_)
print (KNeighbors_Classifier_Model.best_estimator_)
print (KNeighbors_Classifier_Model.best_params_)

Fitting 3 folds for each of 80 candidates, totalling 240 fits
0.8601965471447542
KNeighborsClassifier(algorithm='brute', metric='manhattan', n_neighbors=7)
{'weights': 'uniform', 'n_neighbors': 7, 'metric': 'manhattan', 'algorithm': 'brute'}
```

Fig. 4.30 Codes to perform a random search

times. Of course, you might ask how to ascertain the number of iterations. If you enter a large number such as 1000, you might get a user warning as "`The total space of parameters 80 is smaller than n_iter=1000. Running 80 iterations`". So, this gives you a hint that the maximum iteration you can set for this data is 80.

When you implement these hyperparameters on the model as below, it will look as (Fig. 4.30).

If you compare the outputs from the above methods, you will realise that both are the same. We found that the optimum suggested hyperparameters are as follows:

```
algorithm='brute', metric='manhattan', n_neighbors=7, weights = 'uniform'
```

Bonus: As you might have already noticed that there are two approaches to performing machine learning (supervised): one is through train/test split approach (as we did during our non-leaky code in the previous sections), and the second is using cross-validation (as we did in this section). Now time to compare the outputs of both

```
In [16]: #Step 1: Import libraries, packages and modules
import pandas as pd
import matplotlib.pyplot as plt
from sklearn.model_selection import train_test_split
from sklearn import model_selection
from sklearn.preprocessing import MinMaxScaler
from sklearn.neighbors import KNeighborsClassifier
from sklearn.pipeline import Pipeline

#Step 2: Load the dataset
data = pd.read_csv("C:\\Users\\ghosh\\Downloads\\Pre-processed dataset.csv", header=0, index_col=0)

#Step 3: Assign the target (dependent) variable and the predictor variables
variable_names = ["Binary_Gender_Male", "Age", "Not being able to stop or control worrying", "Little interest or pleasure in doi
X = data[variable_names]     #predictors
y = data["Binary_Feeling nervous, anxious or on edge__Low"] #predicted
X_train, X_test, y_train, y_test = train_test_split(X, y, test_size=0.30, random_state=0)

#Step 4: Build the prediction models
models = []
models.append(('K Neighbors Classifier', KNeighborsClassifier(algorithm='brute', metric='manhattan', n_neighbors=7, weights = 'ur

#Step 5: Evaluate all the models to compare and choose the best
targets = []
algorithm_names = []

for ML_models, model in models:
    kfold = model_selection.KFold(n_splits=10, shuffle=True, random_state=0)
    Variable_Scaler = MinMaxScaler()
    pipeline = Pipeline(steps=[('VariableScaler', Variable_Scaler), ('Model', model)])
    cv_results = model_selection.cross_val_score(model, X, y, cv=kfold, scoring='f1_weighted') #Other scoring metrics available
    targets.append(cv_results)
    algorithm_names.append(ML_models)
    output = "%s: %f (%f)" % (ML_models, cv_results.mean(), cv_results.std())
    print(output)

K Neighbors Classifier: 0.849173 (0.034725)
```

Fig. 4.31 Codes to do machine learning using the cross-validation method (instead of using the train-test split method)

of them. When we utilise our knowledge about hyperparameters on the model (from the above outputs), then it looks as shown in Fig. 4.31.

While using the cross-validation method (refer to the codes in Fig. 4.31).

While Using Train/Test Split Method (Refer to the Codes in Fig. 4.32)

Conclusion

So far, we have been predicting between two categories, such as true or false and yes or no, using two-class classification. It is useful for answering questions such as:

- Is this narrative positive?
- Which of the two advertisements draws more customers?

But then there are times when you might want to ask more complex questions such as:

- What is the mood of this narrative?
- Which of several advertisements draws more customers?

In such instances, you will have more than two categories. That is possible too, and we call that multi-class classification.

```
In [8]: #Step 1: Import all the relevant modules, libraries and packages
import pandas as pd
from sklearn.model_selection import train_test_split
from sklearn.neighbors import KNeighborsClassifier
from sklearn.metrics import confusion_matrix
from sklearn.metrics import accuracy_score
from sklearn import metrics
from sklearn.metrics import roc_auc_score
from yellowbrick.classifier import ClassificationReport
import seaborn as sns
from sklearn.preprocessing import MinMaxScaler

#Step 2: Load the dataset
data = pd.read_csv("C:\\Users\\ghosh\\Downloads\\Pre-processed dataset.csv", header=0, index_col=0)

#Step 3: Assign the target (dependent) variable and the predictor variables
variable_names = ["Binary_Gender_Male", "Age", "Not being able to stop or control worrying", "Little interest or pleasure in doi
X = data[variable_names]     #predictors
y = data["Binary_Feeling nervous, anxious or on edge__Low"] #predicted

#Step 4: Splitting the data into test and train dataset
X_train, X_test, y_train, y_test = train_test_split(X, y, test_size = 0.3, random_state = 0)

#Step 5: Re-scale the variables
Variable_Scaler = MinMaxScaler()
Variable_Scaler.fit(X_train)
X_train = Variable_Scaler.transform(X_train)
X_test = Variable_Scaler.transform(X_test)

#Step 6: Build the prediction classifer model
classifier_model=KNeighborsClassifier()
classifier_model=classifier_model.fit(X_train, y_train)
y_pred = classifier_model.predict(X_test)

#Step 7: Display the performance on the testing set
print ('Accuracy Score :',accuracy_score(y_test, y_pred))
print (metrics.classification_report(y_test, y_pred))
print ('roc_auc_score: ', roc_auc_score(y_test, y_pred))

Accuracy Score : 0.8695652173913043
              precision    recall  f1-score   support

           0       0.90      0.93      0.91       241
           1       0.76      0.70      0.73        81

    accuracy                           0.87       322
   macro avg       0.83      0.81      0.82       322
weighted avg       0.87      0.87      0.87       322

roc_auc_score:  0.8145074535116029
```

Fig. 4.32 Codes to do machine learning using train-test split method (alternative to cross-validation method)

4.3 Regression

4.3.1 Regression Using Machine Learning and How to Interpret the Results

In the previous section, we learnt about predicting categories (whether the person will attempt suicide). In this section, let us learn to predict continuous data. For example, how to predict the frequency of *not being able to stop or control worrying?*

So, a regression can help you predict such frequencies. Note the codes in Fig. 4.33. In step 3, unlike the previous exercises, you will see I have added age (continuous variable) as a predictor variable (along with the other three symptom variables) and set the target variable as the frequency of "Feeling down, depressed or hopeless". In the previous section, we were trying to predict a categorical variable, that is, suicide attempt.

The result will be as shown in Figs. 4.34, 4.35 and 4.36.

Typically, you should expect the *R2* scores to be between 0 and 1, where 1 indicates the best possible model. But as you can see in Fig. 4.34, by running this model

```
In [18]: #Step 1: Import Libraries
from sklearn.ensemble import RandomForestRegressor
from sklearn.metrics import mean_squared_error
from sklearn.metrics import mean_absolute_error
from sklearn.model_selection import train_test_split
from sklearn.metrics import r2_score
import pandas as pd
from math import sqrt

#Step 2: Load the dataset
data = pd.read_csv("C:\\Users\\ghosh\\Downloads\\Pre-processed dataset.csv", header=0, index_col=0)

#Step 3: Assign the target (dependent) variable and the predictor variables
variable_names = ["Age", "Little interest or pleasure in doing things", "Feeling down, depressed, or hopeless", "Thoughts that yo
X = data[variable_names]      #predictors
y = data["Not being able to stop or control worrying"] #predicted

#Step 4: Splitting the data into test and train dataset
X_train, X_test, y_train, y_test = train_test_split(X, y, test_size=0.30)

#Step 5: Re-scale the variables
Variable_Scaler = MinMaxScaler()
Variable_Scaler.fit(X_train)
X_train = Variable_Scaler.transform(X_train)
X_test = Variable_Scaler.transform(X_test)

#Step 6: Build the prediction regressor model
RandomForestModel = RandomForestRegressor()
RandomForestModel.fit(X_train, y_train)
ypred = RandomForestModel.predict(X_test)

#Step 7: Estimate the performance of the RandomForestRegressor for predicting continuous variables
score = RandomForestModel.score(X_train, y_train)
print("R2_Score_train dataset: ", score)
print("R2_Score_test dataset: ", r2_score(y_test, y_pred))
mse = mean_squared_error(y_test, y_pred)
print("MSE: ", mse) #Mean Squared Error (MSE)
print("RMSE: ", sqrt(mse)) #Root Mean Squared Error (RMSE)
print ("MAE: ", mean_absolute_error(y_test, y_pred)) #Mean Absolute Error (MAE)

#Step 8: Create a Residuals Plot
from yellowbrick.regressor import ResidualsPlot
visualiser_Residuals_Plot = ResidualsPlot(RandomForestModel, hist=True)
visualiser_Residuals_Plot.fit(X_train, y_train)
visualiser_Residuals_Plot.score(X_test, y_test)
visualiser_Residuals_Plot.show()
r = y_test-y_pred
print ("residual:",r)

#Step 9: Create a Prediction Error Plot (to visualise the variance in the model)
from yellowbrick.regressor import PredictionError
visualiser_prediction_error = PredictionError(RandomForestModel)
visualiser_prediction_error.fit(X_train, y_train)
visualiser_prediction_error.score(X_test, y_test)
visualiser_prediction_error.show()
```

Fig. 4.33 Codes for conducting regression with machine learning

```
R2_Score_train dataset:  0.6725667310214103
R2_Score_test dataset:  -1.4411162495145975
MSE:  2.6677018633540373
RMSE:  1.6333100940586993
MAE:  1.2826086956521738
```

Fig. 4.34 Result of the regression

on the sample data, we found the R2 score of the test data set in the negative (i.e. −1.44) and then the chosen model fits worse than a horizontal line (and so the *R*2 is negative).

The **R-square** (coefficient of determination) is the proportion of variance (of y) that the independent variables in the model have explained. A normal R^2 value should fall between 0 and 1; the higher R^2, the better. The training performance ($R^2 = 0.93$) is better than the testing performance ($R^2 = 0.20$). I know the results on the test data are poor, and that's what will happen if your variables are not relevant

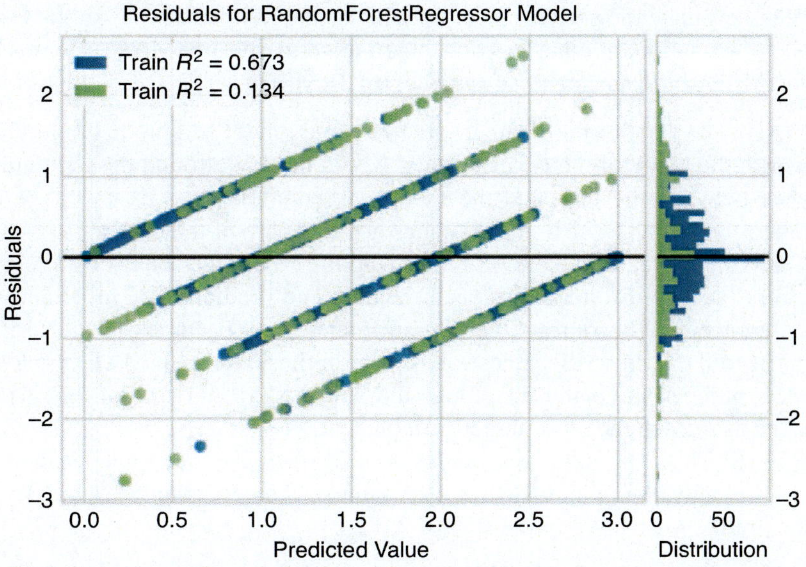

Fig. 4.35 Scatter plot of the residuals for the regressor model

Fig. 4.36 Plot for the prediction error for regressor

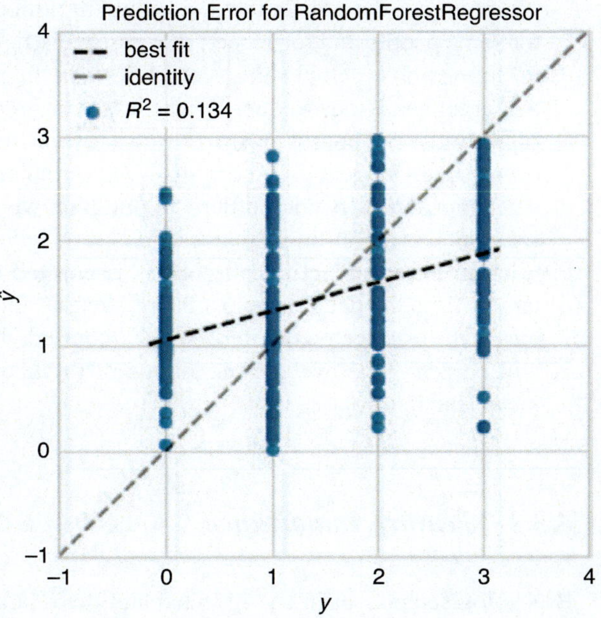

to predicting the target. Suppose you get such poor results, then time to revaluate which variables have you considered and refer to the literature for inspiration. Other times you would want to check if you have properly cleaned and formatted the dataset.

But to evaluate the performance, we will use the R-square value **on the test data**.

Next, we will learn other metrics: m**ean absolute error** (MAE), **mean squared error** (MSE) and **root mean squared error** (RMSE).

- **MAE:** The smaller the number, the better the model is a prediction. In an ideal world, the MAE of the perfect model would be 0. We found the absolute difference between the data and the models' predictions (MAE) is 0.15. Our data ranges between 0 and 1, and this does not take care of the outliers.
- **MSE:** In this case, we assign more weight to the larger errors by squaring the errors (i.e. the difference between the data and prediction). This implies that a difference of 2 becomes 4, and the difference of 4 becomes 16. Our MSE was equal to 0.04, and MSE takes care of the outliers. As with MAE, the lower the MSE number, the better it is. Note, unlike the MAE, the unit of MSE is not the same as the dependent or target variable.
- **RMSE:** This is simply the square root of MSE. It also takes care of the outliers. Additionally, the unit of RMSE is the same as the dependent variable (like MAE and unlike MSE). We found RMSE = 0.20.

Residual plot (= observed value - predicted value): The residual plot helps us analyse the error variance of the regressor. We plotted the difference between the expected and actual values in the above case. There is a residual value for each person in our dataset. In the plot, you will notice that the residual was negative for some people and for others; it is positive. The residual will be in NEGATIVE for those whose actual value was LESS than the predicted value. This is because the target behaviour frequency was lesser than your model predicted. Likewise, suppose the residual is positive for a person. In that case, the target behaviour occurrence frequency is more than what your model predicted. It can help us detect non-linearity and outliers. In our case, we are using random forest regressor, which is a non-linear regressor.

Prediction error (= actual outcome - predicted outcome): It shows the actual targets from the dataset against the predicted values generated by our model. The bias is the average prediction error. If unbiased, then the average prediction error will be zero. The average magnitude of prediction error is called RMSE (root mean square error).

4.3.2 Feature Importance

Psychologists have been trying to explain the relationships between different phenomena and individual experiences. For example, if you were interested in studying stress resilience in students, you would probably be curious to know the identifying predictors of resilience. You might consider studying a range of potential predictor variables such as the locus of control scale, personality factors and symptoms of anxiety and depression and collect the data using the survey method. The next part consists of pre-processing the data. Once that is done, then we will jump into analysing the data.

Standard statistical method: We will perform multiple regression to find which variables are significantly related to resilience.

Alternative machine learning method: OK – we learnt how to build machine learning models and report the accuracy of the predictions. But then what? How do we use it in our research project? What is the next step?

This is an obvious question, and this lesson hopes to address how to use those machine learning algorithms that we just built. Yes, I am talking of both classifiers and regressors.

Well, there are two routes:

1. We can deploy the built machine learning models as applications. There will be a place to enter the input values, throwing out an output value. This is what data/computer scientists in the industry do. They build machine learning models to boost their web or mobile phone applications. This is why they are more interested in the accuracy level of the output. Take the example of a data scientist working with a team that runs a movie streaming app. The person's job is to improve the app users' experience by predicting and recommending which movie or show a particular user might like to watch next based on what they had watched before.

   I would assume that it matters little to this person to answer the WHY. There is a good rationale for that. The top priority of the computer scientist or the data scientist in that business team is probably user experience, making their users happy and more engaged with their content and not explaining or understanding why a user watched this movie instead of others. Neither their customers nor users care what is predicting why they watched this movie instead of others. Instead, the customer will pay them or stay more engaged if the app accurately predicts what they want to watch next. So, as you might have imagined, the practitioners of machine learning in business are often more concerned about the prediction accuracy of their model. They want to find out the most important features and engage in dimensionality reduction procedures. But even then, they aim to improve the prediction performances of the algorithm. I would love to teach you to create a recommender system like Netflix or a web app to predict a target behaviour. Given the introductory nature of this book, let us save it for a more advanced book. Maybe, I will start writing a book focused more on the industrial application of behavioural science.
2. But we are psychologists, and our interest differs. Behavioural and social scientists are often intrigued by the question of what determines a target behaviour. There might be a dozen of input variables in the dataset, and we have a column for target behaviour. But how do we know which variables predicted the occurrence of the target behaviour – and to what extent. We are more interested in the WHY questions. Often we want to know which independent variables were the biggest contributor to the prediction. Why a person performs a particular (target) behaviour? So, it matters to us which variables predicted our target behaviour

and to what extent. The resultant knowledge might be utilised to build behavioural models or justify their findings. This can be made possible by feature importance functionality. In this section, we will learn about it.

Introduction to the Black Box Phenomena

Suppose you have a piece of background knowledge in statistics. In that case, you might be wondering, "but we do get the predictor variables which are statistically significant from their p-values!". That is not the case in machine learning. Many of the machine learning models are "black boxes". In the simplest possible sense, it means we know what data goes in and what decision comes out. But the processes between input and output are often not visible. While such explanations are not a big concern for simple models such as decision trees or linear regression, for complex models, such as deep neural networks, black box phenomena are more prominent. Therefore, such complex machine learning models are often infamous for being not transparent about their decision-making process. The high accuracy of the model is not a sufficient condition to build trust among users. It is sometimes important to consider how the model made its decision to ensure safety, fairness and trustworthiness. Feature importance helps us look inside the black box. It enables us to see which variables were leveraged and to what extent to generate the prediction. In relation to this, let me touch upon a very important issue related to biases in machine learning decisions.

Machine learning is increasingly used to make important clinical judgements (e.g. Meyer et al., 2018), predict judicial decisions (e.g. Medvedeva et al., 2019) and decide whether to sanction loans in several other ways in which it can impact our daily life. For example, imagine an algorithm that predicts who's guilty of a crime based on one's race or utilises gender to make recruitment decisions. This could lead to biased decision-making and might affect the lives of millions. Therefore, as a data science practitioner, I would urge you to be mindful of these potential issues that might creep up due to bad data. For your aid, different authorities are publishing their guidelines. Following them might help the readers of your research paper gain trust in your findings. One example is that of the UK National Health Service (NHS). So, suppose you are working in the healthcare domain. In that case, you can follow the UK NHS's initial code of conduct for data-driven health and care technology (NHS, 2021), especially if your country does not yet have its own code of conduct. Interestingly, although machine learning is increasingly being deployed to assist physicians' decision-making, it is important to note the word "assist" and not a replacement. The decision's liability would remain with the clinician (at least, it's true for the year 2021).

One simple strategy to **avoid biases in the data** is to remove the sociodemographic variables such as race or gender whenever possible – right at the beginning of the data pre-processing stage. This is more doable for some studies than others. For example, if we are trying to understand what predicts productive workplace behaviour, it would probably be unfair to include the race and gender of a person in the prediction model. In such cases, especially if the literature supports those sociodemographic variables of a person are not related to a particular behaviour, it should not be included in the dataset.

```
In [45]: #import libraries, packages and modules
         import pandas as pd
         import numpy as np
         import matplotlib.pyplot as plt
         from sklearn.ensemble import RandomForestRegressor
         import shap

         #Load the dataset
         data = pd.read_csv("C:\\Users\\ghosh\\Downloads\\Pre-processed dataset.csv", header=0, index_col=0)

         #Assign the target (dependent) variable and the predictor variables
         variable_names = ["Binary_Gender_Male", "Age", "Not being able to stop or control worrying", "Little interest or pleasure in doi
         Predictors = data[variable_names]     #predictors
         Target = data["Thoughts that you would be better off dead or of hurting yourself in some way"] #predicted

         #Building the model
         RF_forest = RandomForestRegressor().fit(Predictors, Target)
         RF_forest.feature_importances_

Out[45]: array([0.05953956, 0.40462806, 0.12142036, 0.10951922, 0.2658141 ,
                0.0390787 ])
```

Fig. 4.37 Codes to compute feature importance for a regressor (using the built-in feature importance function)

If you are unsure whether to exclude sociodemographic variables for a more complicated research topic, and even when we exclude the sociodemographics, which of the remaining independent variables contribute to the decision, then feature importance will help open the black box.

Feature importance explains the contributory relationships between independent variables and dependent variables. In our case, we had been predicting variable 14 from variables 11, 12 and 13. It's great that we found an accurately performing machine learning model in the previous section. But that is not enough. We need to understand which independent variables influenced the most and the least the prediction out of the three. Again there are many ways to estimate feature importance. But I will show two major ones. Although I demonstrate feature importance for regressor in this section, you can replace the regressor model with your classifier and still get your result.

The simplest way to compute feature importance for a random forest model is probably to opt for the **tree-specific feature importance measure.** It estimates the average reduction in impurity across all trees due to each feature. So, using this built-in feature importance function of the random forest regressor, we can compute the independent variables' feature importance, as shown in Fig. 4.37.

The result found was as follows:

- Binary_Gender_Male: 0.05953956
- Age: 0.40462806
- Not being able to stop or control worrying: 0.12142036
- Little interest or pleasure in doing things: 0.10951922
- Feeling down, depressed or hopeless: 0.2658141
- Binary_Feeling nervous, anxious or on edge__Low: 0.0390787

This indicates that *Age* has the highest contribution or importance in predicting *Thoughts that you would be better off dead or of hurting yourself in some way* (the dependent variable).

Explaining the Model with SHAP Values (Lundberg & Lee, 2017)

Another way of finding the feature importance of independent variables is using SHAP (SHapley Additive exPlanations). Personally, I love this method. As above, features with large absolute Shapley values are more important than others. SHAP values are based on the magnitude of feature attributions and are a method from coalitional game theory.

Accordingly, we can see below that variable 13 has the strongest influence, followed by variables 11 and 12. We found exactly the same result (in terms of ranking) using the previous method. But I personally prefer using SHAP values because it has a solid theoretical foundation in game theory and arguably distributes the effects fairly. As mentioned above, the black box phenomenon haunts the complex models such as artificial neural networks more than the simpler ones such as linear regression. But there is still a reason why I use SHAP values for linear regression. Usually, in linear regression, as you already know from statistics classes, we typically examine the coefficients learnt for each feature. Those coefficients tell us what will happen when we change the value of an input feature. But each coefficient's value depends on the input features' scale. So these coefficients are not very good at measuring the overall importance of a feature. In short, the magnitude of a coefficient is not necessarily the best estimate of a feature's importance in a linear regression model (Fig. 4.38).

The result will appear as shown in Figs. 4.39, 4.40 and 4.41.

In Fig. 4.40, each point on the plot represents Shapley value (on the x-axis) for a feature and an instance. The colour represents the value from high to low. Together, they represent the relationship between the value of a feature and the impact on the prediction.

```
In [32]: #import libraries, packages and modules
import pandas as pd
import numpy as np
import matplotlib.pyplot as plt
from sklearn.ensemble import RandomForestRegressor
import shap

#Load the dataset
data = pd.read_csv("C:\\Users\\ghosh\\Downloads\\Pre-processed dataset.csv", header=0, index_col=0)

#Assign the target (dependent) variable and the predictor variables
variable_names = ["Binary_Gender_Male", "Age", "Not being able to stop or control worrying", "Little interest or pleasure in doi
Predictors = data[variable_names]     #predictors
Target = data["Thoughts that you would be better off dead or of hurting yourself in some way"] #predicted

#Building the model
RF_forest = RandomForestRegressor(max_features = 2, random_state=0).fit(Predictors, Target)

explainer = shap.TreeExplainer(RF_forest)
shap_values = explainer.shap_values(Predictors)
shap.force_plot(explainer.expected_value, shap_values)
shap.summary_plot(shap_values, Predictors, plot_type="bar")
shap.summary_plot(shap_values, Predictors)
shap_sum = np.abs(shap_values).mean(axis=0)
importance_data = pd.DataFrame([Predictors.columns.tolist(), shap_sum.tolist()]).T
importance_data.columns = ['feature_name', 'shap_importance']
importance_data = importance_data.sort_values('shap_importance', ascending=False)
print (importance_data)

#importance_data.to_excel('You can name this file name anything.xlsx', engine='xlsxwriter')
```

Fig. 4.38 Codes to compute feature importance for a regressor (using SHAP values)

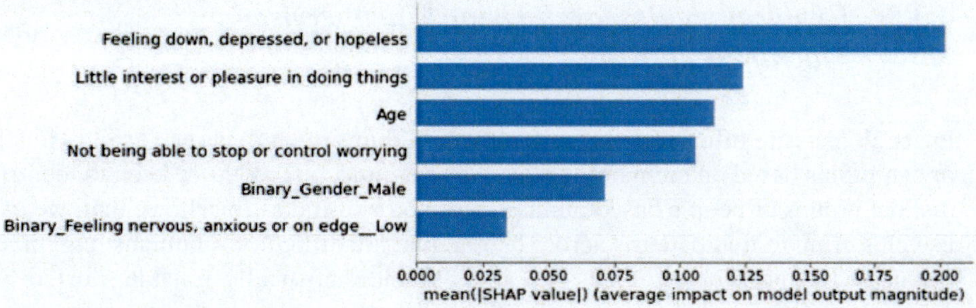

Fig. 4.39 Bar graph showing which features are more important towards making the prediction than others

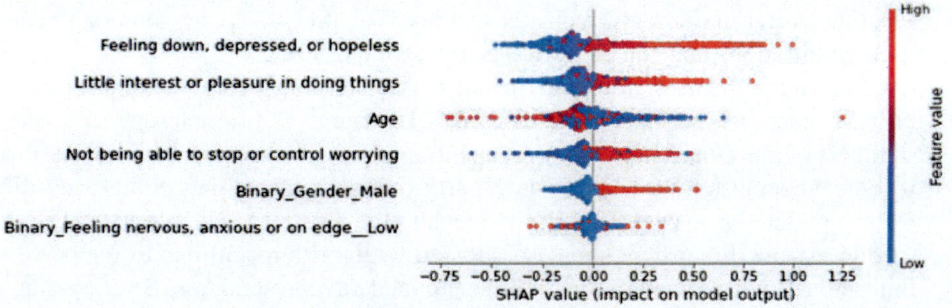

Fig. 4.40 Plot demonstrating the SHAP value for each feature

	feature_name	shap_importance
4	Feeling down, depressed, or hopeless	0.202188
3	Little interest or pleasure in doing things	0.124223
1	Age	0.113081
2	Not being able to stop or control worrying	0.105732
0	Binary_Gender_Male	0.071408
5	Binary_Feeling nervous, anxious or on edge__Low	0.034106

Fig. 4.41 Table demonstrating the SHAP importance value (numerically) for each feature

Endnote

Now you can write an entire result and discussion section explaining or arguing (from the literature) WHY variable 13 had the highest contribution to predicting variable 14. This way, feature importance allows us to open the black box of machine learning. Outside explaining a behaviour, explainability is important in real-world industrial applications too. For example, imagine the machine learning algorithm predicting the best course of treatment for a patient. It's great that the machine suggests something. But at the same time, the physician must understand which independent variables were prioritised while making the prediction. This is why feature importance is growing in demand and recognised importance in machine learning literature.

4.3.3 Exploratory Research Using Unsupervised Machine Learning

Psychologists are often curious to identify subgroups of entities such as individuals or symptoms based on their similarities. For example, one might be keen to identify distinct groups of people based on their symptoms or traits. To achieve that, we can use clustering to find patterns across people that are difficult for humans (e.g. practitioners) to find. For example, one study found statistically reliable clusters of depressive symptoms (Dwyer et al., 2018). Furthermore, the study reported that these clusters differed in their responsiveness to treatment both within and across different antidepressant medications. Now, as you can imagine that such algorithms have massive real-world implications. For example, the above finding can help us select the best drug for a given cluster that may benefit better. Can you imagine any research questions that can be answered by such clustering?

Clustering is the most important unsupervised learning technique. I hope you got the basic idea behind clustering: to reveal subgroups within heterogeneous data. Members of one cluster are more homogenous among themselves than the members of some other cluster. We perform such partitioning on a set of data points according to some prefixed measure of similarity (such as the distance), which we will learn in a while. As the dimensions increase, clustering algorithms fail due to the curse of dimensionality. One needs a lower-dimensional data representation. So clustering is often preceded by dimensionality reduction. Usually less than 10. The reduction could be manual or using any of the usual methods such as PCA.

Note that we have been learning supervised machine learning, where we have a target variable so far. We were trying to build a model with independent variables to predict it. But from this lesson onwards, we will focus on unsupervised machine learning where we do not have to specify any target variables. We need a set of variables based on which the similarity between members or entities shall be estimated, and accordingly, clusters will be formed. In this section, we will attempt to group individuals based on the similarity of their responses to "Not being able to stop or control worrying", "Little interest or pleasure in doing things" and "Feeling down, depressed, or hopeless".

4.4 Clustering

4.4.1 Hierarchical Clustering

Schizophrenia is a heterogeneous condition. One assumption to reduce the heterogeneity is to create subgroups. In 1983, a study attempted to achieve that. It considered patients diagnosed with schizophrenia. Each patient was rated on a checklist for the presence or absence of 42 items extracted from six commonly used operational criteria for schizophrenia (Farmer et al., 1983). The study then used both the complete linkage hierarchical method and the k-means algorithm to generate clusters. Both methods used Euclidean distance as their measure of similarity. The study found two reasonably distinct groups. The first group/cluster was larger and was characterised by good premorbid adjustment, later onset and well-organised delusions.

On the other hand, the second group of patients was characterised by bizarre behaviour and incoherent speech, blunted affect, early onset, poor premorbid adjustment and a family history of schizophrenia. Thus, the hierarchical and k-means clustering methods generated similar profiles. This was to give you an idea of how you might utilise such clustering methods to solve these kinds of problems in psychology. So, as we just noted, there are two types of clustering (among the most used ones): flat and hierarchical. K-means clustering is an example of flat clustering. In this section, we will learn about hierarchical clustering. In the next section, we will learn about k-means clustering.

If you recall, back in your childhood days, you learnt about the classification or taxonomy of the animal kingdom using a diagram like this. Or maybe you have seen your family tree. It probably looked somewhat as shown in Fig. 4.42.

The output of hierarchical clustering is similar to this. For example, let us consider a small dataset of ten respondents, as shown in Fig. 4.43.

We aim to group similar individuals based on their similarity in symptomatic responses. The output for this would look as follows (Fig. 4.44):

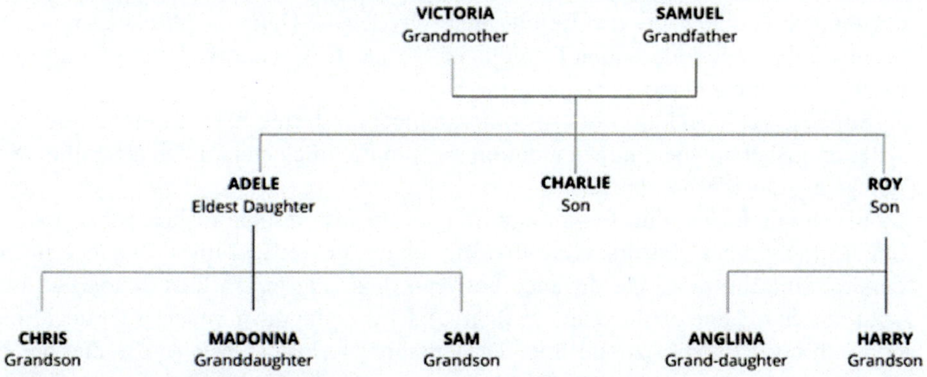

Fig. 4.42 An example of a family tree

Names	Feeling nervous, anxious or on edge	Not being able to stop or control worrying	Worrying too much about different things	Trouble relaxing	Being so restless that it is hard to sit still	Becoming easily annoyed or irritable	Feeling afraid as if something awful might happen	Little interest or pleasure in doing things	Feeling down, depressed, or hopeless	Trouble falling or staying asleep or sleeping too much	Feeling tired or having little energy	Poor appetite or overeating	Feeling bad about yourself – or that you are a failure or have let yourself or your family down	Trouble concentrating on things, such as reading the newspaper or watching television	Moving or speaking so slowly that other people could have noticed? Or the opposite – being so fidgety or restless that you have been moving around a lot more than usual	Thoughts that you would be better off dead or of hurting yourself in some way
Harry	2	2	2	2	0	0	2	1	[illegible]	0	1	0	0	0	0	0
Byron	3	2	1	3	1	1	3	2	1	1	3	3	2	3	0	1
Duncan	0	0	0	0	0	0	0	0	0	0	0	2	0	0	0	0
Ron	3	3	3	3	3	3	3	3	3	3	3	3	1	3	2	0
Sam	1	0	1	1	0	0	0	0	0	2	1	0	0	0	0	0
Heena	2	2	2	2	1	2	2	2	1	2	2	2	1	2	0	0
Pooja	1	0	1	1	0	0	1	0	0	1	1	3	1	2	0	0
Daniella	3	3	2	2	0	2	2	2	2	1	0	0	1	3	0	0
Roshan	2	1	1	1	1	1	1	1	1	1	1	1	0	1	0	0
Natalie	3	3	3	3	3	3	3	3	3	3	3	3	3	3	3	0

Fig. 4.43 Sample dataset

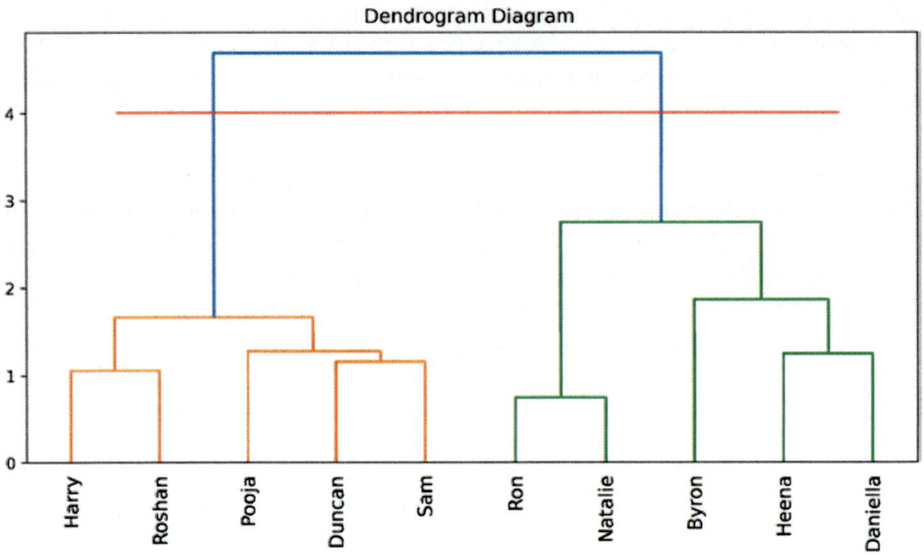

Fig. 4.44. Example of a dendrogram

How to Interpret Dendrogram?

The x-axis consists of the survey respondents. The y-axis consists of the Euclidean distance between the groups or clusters. Each of these people is being paired into clusters as we move up at different hierarchy levels. Harry and Roshan reported more similar experiences than Daniella or Heena. This is similar to the family tree example mentioned above.

Suppose you check the codes for hierarchical clustering below. In that case, you will notice that nowhere do we mention the number of clusters in the algorithm. So, how many clusters are there?

It's simple to find out. First, determine the largest vertical distance that doesn't intersect any other clusters. Note that the longer the vertical lines' distance in the dendrogram, the more the distance between those clusters. I have demarked this with a thick red line on the diagram (above). Now count the number of vertical lines going through this horizontal line. The number of clusters will be the number of vertical lines intersected by the line drawn using the threshold. Can you see the two vertical lines?

So, there are two clusters for this dataset. Although sometimes the longest vertical distances are not easy, there could be more depending on where you cut off.

Great. We created a dendrogram for agglomerative clustering, and now we know how to infer the diagram.

Other Possible Use Cases We used hierarchical clustering for grouping. You can group people, different activities, traits, products, interventions, psychopathological symptoms, items of a questionnaire and the like. Possibilities are limitless. You can study any construct. The only condition is that you should be able to break the construct down into constitutive particles. In the case of the example above, we tried to group people based on their similarities in symptomatic experiences. So, the construct was made up of 16 symptoms. But you can use any variable of your choice.

These are all good examples of how we can structure data in a tree-like fashion and use it to detect groups or clusters in the data. Now answer me this. Looking at the family tree above, who do you think Harry is most genetically similar to? Out of Anglina and Madonna?

If you think it's Anglina then you are absolutely correct. Harry and Anglina have a common father. Therefore, they are most similar to each other than their cousin Madonna. Madonna is more similar to Sam and Chris. In terms of clusters, then Chris, Madonna and Sam are in one cluster. Anglina and Harry are in another cluster.

Note that this is not the best example of hierarchical clustering because no clustering algorithm was used to make this family tree. But suppose we had collected genetic materials from our family members and ran the hierarchical algorithm for genetic similarities on the dataset. In that case, chances are we would get a figure like this. So the family tree above is just an example to show the idea of clustering based on similarities. There is much more information offered by clustering diagrams (also known as dendrograms), as we will find soon.

There are two types of hierarchical clustering. divisive and agglomerative. First, let us learn about agglomerative in five easy steps:

Step 1: You collected genetic materials from Chris, Madonna, Sam, Anglina and Harry. Consider each of these five people as a cluster. So, there are five clusters right now.

Step 2: In the next step, the algorithm will attempt to build a bigger cluster by joining the two closest data points. So, in this step, the number of clusters became two (for Adele and Roy).

Ignore Charlie because we assumed he was dead and had no offspring.

Step 3: Finally, in this step, the two clusters were fused to form one.

Did you notice that we started with five small clusters and reduced them to one big cluster? This is what we do in agglomerative hierarchical clustering. On the other hand, divisive clustering is just the opposite. It starts at one cluster, and splits are performed as one moves down the hierarchy. Hence, it is a "top-down" approach.

We visualise the result in the form of a dendrogram. It looks quite similar to the family tree above. The dendrogram shows the hierarchical relationship between objects (or people, as in the above example). It allows you to figure out the best possible way to allocate objects (or people) to clusters.

You need to know two facts about the dendrogram diagram to interpret it. First, the horizontal axis represents the distance (or dissimilarity) between clusters. Second, the vertical axis represents the objects (or people) and clusters. These two facts will help you determine the next important question.

How Do We Find the Optimal Number of Clusters from a Dendrogram? It's simple. We look for the largest vertical distance in the diagram (that passes without crossing any horizontal line). In the above diagram, look at the red-framed line. Let's count the number of lines on the diagram and determine the optimal number of clusters.

Let us dive in with some real-world problems. You will gain more clarity after seeing hierarchical clustering in practice. If you have studied mental health, you are familiar with the *Diagnostic and Statistical Manual of Mental Disorders* (DSM). If not, then do not worry. It is simply the diagnostic manual published by the American Psychiatric Association. The DSM proposes different categories of mental disorders and which symptoms should come under each of those disorders. **Quick fact:** Recent studies show that DSM has low reliability and validity and is potentially unfit for clinical use. There are some serious problems with it, and further discussion is beyond the scope of this book. But you need to know here that lack of empirical basis was one of the issues. Right now, our goal is to learn hierarchical clustering. So, let us imagine that we are trying to rebuild our own version of typologies in mental disorders.

We will use the dataset which has item-wise responses from the 1958 community sample on the Generalised Anxiety Disorder scale (GAD-7; Spitzer et al., 2006) and the Patient Health Questionnaire (PHQ-9; Kroenke et al., 2001). For the tutorial purpose, let's pick three variables: one symptom of anxiety and two symptoms of depression. We have pre-processed it (using the steps taught in the previous section), so we will not redo it. Still, we shall attempt to group the remaining people in the sample (after pre-processing) based on their self-report similarities. Such grouping or clustering is expected to help understand which people are similar in terms of their symptomatic experiences. This can aid in treatment planning and

```
In [22]: #Step 1: Import libraries
import pandas as pd
import scipy.cluster.hierarchy as cluster
import matplotlib.pyplot as plt
from sklearn.preprocessing import normalize
from sklearn.cluster import AgglomerativeClustering

#Step 2: Load the dataset
data = pd.read_csv("C:\\Users\\ghosh\\Downloads\\Pre-processed dataset.csv", header=0, index_col=0)
variable_names = ["Not being able to stop or control worrying", "Little interest or pleasure in doing things", "Feeling down, de
X = data[variable_names]

#Step 3: Normalise the data
scaler = MinMaxScaler()
normalised_data = pd.DataFrame(scaler.fit_transform(X))
normalised_data.columns = ["Not being able to stop or control worrying", "Little interest or pleasure in doing things", "Feeling
normalised_data.describe()


#Step 4: Plot the dendrogram
plt.figure(figsize=(10, 7))
plt.title("Dendrogram Demo")
dend = cluster.dendrogram(cluster.linkage(normalised_data, method='ward'), leaf_rotation=90)

#Step 5: Create an instance of Agglomerative-Clustering
modelx = AgglomerativeClustering(affinity='euclidean', linkage='ward')
modelx.fit(normalised_data)

#Step 6: Return an array of integers corresponding to each categories or clusters
#print (modelx.labels_)

#plt.savefig('plt.png', dpi=520, format='png', bbox_inches='tight')
```

Fig. 4.45 Code for hierarchical clustering

policymaking. Note, this is just an example for the tutorial. The aim is to make clusters from this data that can segment similar patients together. The codes for performing hierarchical clustering are shown in Fig. 4.45.

After we have imported all the libraries and loaded the data, we will need to change the numeric values in the dataset to a common scale. For that purpose, we will rescale the data. For example, if you had "age" and "income" in your dataset, the range for both parameters is different. While age ranges from 0 to 100 (approx.), the minimum income for a single person (without dependents) starts from £20,383 in the United Kingdom. That way, numerically, income is about several times larger than age. In such cases, where your features have different ranges, variables like income with the larger range might appear more influential than age because it has a larger value. That would be misleading.

So, we must rescale such data to bring all the variables to the same range.

NEXT, we build the dendrogram and plot it. While doing so, we would have to set the METHOD, and that would tell us how clusters are separated from each other. The WARD method we are using here is based on the sum of squares. There are other methods, such as the single method, complete method and average method.

The single method is based on two closest objects. The complete method is used when we want to base it on two farthest objects. The average method will base on the arithmetic mean of the objects. The centroid method will base on the geometric mean of all objects. The median method is if you want to base the clustering on the median of all objects. There are a lot of other variations. I encourage you to go

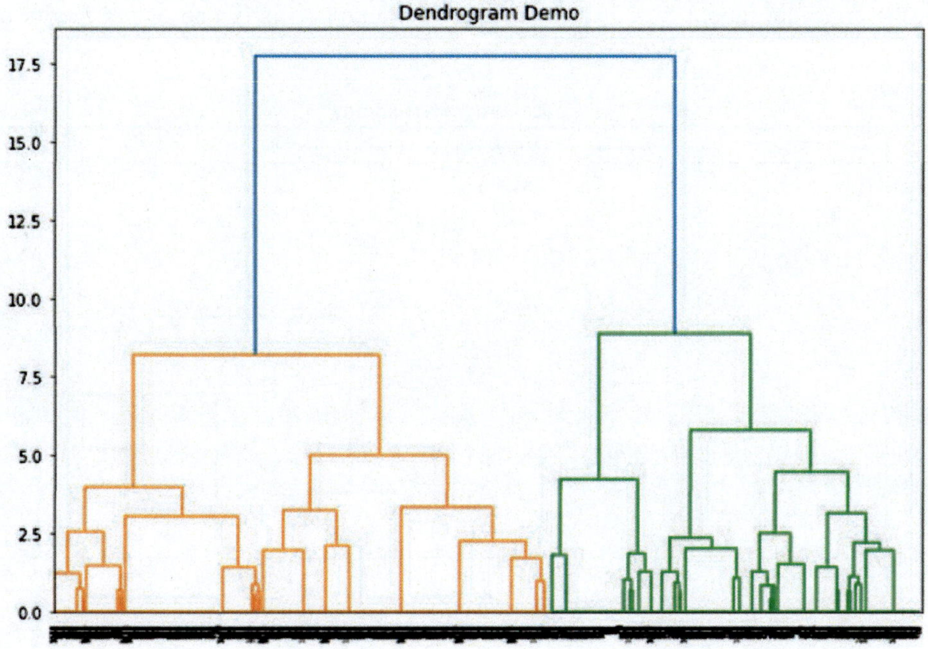

Fig. 4.46 Dendrogram

ahead and try out these different variations. You will find more information about the method: https://scikit-learn.org/stable/modules/generated/sklearn.cluster.AgglomerativeClustering.html

We will also create an instance of `AgglomerativeClustering` using the Euclidean distance to measure the distance between data points and ward linkage. This will calculate the proximity of clusters. The `labels_` property here will return an array of integers where the values correspond to the distinct categories. Now, if you HIT run, Python should generate a dendrogram. A dendrogram shows the hierarchical relationship between similar sets of data.

The output, dendrogram diagram for a large dataset, can be found in Fig. 4.46.

Now, of course, the x-axis which represents the respondent's name or ID is illegible above. But you can always save the plot to an image file with `plt.savefig('sample_image.png')` and then zoom it in. But understandably, manually scanning through so many peoples' names or IDs might be a bother. So, you might wish to use this approach when you have fewer entities to study. For example, if you had ten patients, you might have got a graph, as shown in Fig. 4.47, which shows the names more legibly. Do you see how easy it gets to infer that Harry and Roshan are more similar to each other than Daniella?

That's all for this section. Next, try and experiment with hierarchical clustering with your own data and see how it changes from this graph.

Endnote

While writing the code for hierarchical clustering in this section, we did not inform the algorithm about how many clusters we wanted. This is one of the merits of

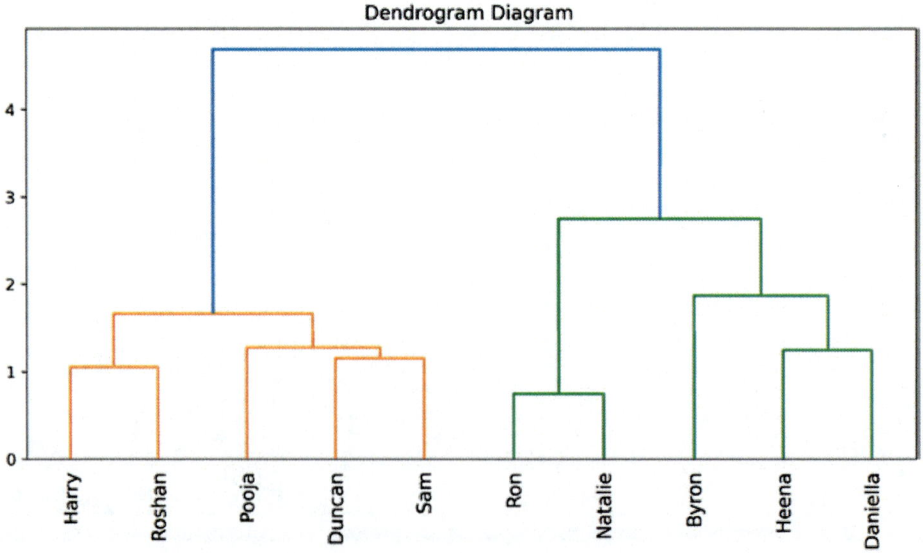

Fig. 4.47 Dendrogram for a dataset with ten people

hierarchical clustering, especially when we do not know how many clusters are in advance. Secondly, hierarchical clustering produces replicable clusters. Given these merits, hierarchical clustering cannot handle big data well, like in cases where:

1. Your dataset is very large.
2. You have a rough idea about the number of clusters you want.
3. Replicability is not a top priority.

We would go for k-means clustering, which we will learn in the next section.

4.4.2 K-Means Clustering

Do you remember the schizophrenia study mentioned at the start of the previous section, which used both hierarchical clustering and k-means clustering? In the previous section, we learnt about hierarchical clustering. In this section, we will learn about k-means clustering.

Suppose we have a dataset about people with certain variables (e.g. symptoms) and values for these variables. You wish to categorise those people into relatively homogenous groups.

Memory Refresher The word "variable" is not used in Python often. In this book, we have used "variable" in a way you did in statistics, and it's for your convenience. But just so that you know, machine learning practitioners (and most books written by them) would often refer to "variable" as "feature". They mean the same. Therefore, when you come across concepts such as feature importance or dimensionality reduction, know that they mean to determine the importance of each predictor variable and intend to reduce the number of variables.

In the previous lesson, we learnt to group people based on their self-reported frequency of anxiety and depressive experiences. Note that there was no dependent variable. All of them were independent variables. It is the same for k-means clustering. Here, the goal is partitioning "n" observations into "k" clusters. Therefore, "k" simply means a number of clusters. We will cover how we estimate it below using the same dataset as above.

The required code is shown in Fig. 4.48.

If you noticed the code carefully, k-means asks you to specify the number of clusters (unlike the hierarchical clustering algorithm) for the given dataset. But how do we do that?

The number of clusters that we choose for k-means cannot be random. But do not worry; there are methods to determine that.

I have demonstrated two methods here:

1. **The elbow method:** Randomly initialises the k-means algorithm for a range of K values. Then plots it against the within-cluster sum of squares (WCSS) for each K value. See below for the elbow point graph. As the number of clusters

```
#Step 1: Import libraries, packages and modules
import pandas as pd
import matplotlib.pyplot as plt
from sklearn.cluster import KMeans
from sklearn import metrics
from sklearn.metrics import silhouette_score
import seaborn as sns
from yellowbrick.cluster import silhouette_visualizer
from sklearn.preprocessing import MinMaxScaler
import numpy as np

#Step 2: Load the dataset
data = pd.read_csv("C:\\Users\\ghosh\\Downloads\\Pre-processed dataset.csv", header=0, index_col=0)
data.head()

#Step 3: Assign the target (dependent) variable and the predictor variables
variable_names = ["Not being able to stop or control worrying", "Little interest or pleasure in doing things", "Feeling down, de
x = data[variable_names]

#Step 4: Normalisation of the dataset (feature scaling) to avoid building biased model.
scaler = MinMaxScaler()
normalised_data = pd.DataFrame(scaler.fit_transform(x))
normalised_data.columns = ["Not being able to stop or control worrying", "Little interest or pleasure in doing things", "Feeling
normalised_data.describe()

#Step 5: Find the optimum number of clusters

#Method 1: Using Elbow Method
SSE = [] #Sum of Squared Errors (SSE)
for k in range(0,10):
    kmeans = KMeans(n_clusters = k+1, init='k-means++', random_state = 1).fit(normalised_data)
    SSE.append(kmeans.inertia_)
sns.pointplot(x=list(range(1,11)), y=SSE).set(title='Elbow point graph', xlabel='No. of clusters', ylabel='Within-Cluster-Sum-of-
plt.show()

#Method 2: Using Silhouette Method
silhouette_visualizer(KMeans(2, random_state =1), normalised_data, colors = 'yellowbrick')
score = silhouette_score(x, kmeans.labels_, metric='euclidean')
print("The Silhouette Score is %.2f" % score)

#Step 6: Clustering with KMeans
kmeans_model = KMeans(n_clusters=2, init='k-means++', random_state= 0).fit(normalised_data)
centers = kmeans_model.cluster_centers_
fig = plt.figure()
ax = fig.add_subplot(111, projection = '3d')
scatterplot_fig = ax.scatter(normalised_data["Not being able to stop or control worrying"], normalised_data["Little interest or 
centers = kmeans_model.cluster_centers_
ax.scatter(centers[:, 0], centers[:, 1], centers[:, 2], c = 'black')
plt.title("3D scatter plot for donor segments")
ax.set_xlabel('X-axis', fontweight ='bold')
ax.set_ylabel('Y-axis', fontweight ='bold')
ax.set_zlabel('Z-axis', fontweight ='bold')
fig.colorbar(scatterplot_fig, ax = ax, shrink = 0.5, aspect = 5)
```

```
#Step 7: Transform the normalised data back to original encoding.
reversed_data_transf = pd.DataFrame(scaler.inverse_transform(normalised_data))
reversed_data_transf.columns = normalised_data.columns
reversed_data_transf['Cluster'] = kmeans_model.labels_
inver_new_data = scaler.inverse_transform(normalised_data)
#print (reversed_data_transf)

#Step 8: Computing cluster centers and predicting cluster index for each sample object.
y = kmeans_model.fit_predict(reversed_data_transf[["Not being able to stop or control worrying","Little interest or pleasure in 
reversed_data_transf['Cluster'] = y
#print(reversed_data_transf.head(10))
reversed_data_transf.to_excel('K-Means_Result_tutorial_book.xlsx', engine='xlsxwriter')

#Step 10: Evaluating the performance of the KMeans
labels = kmeans_model.labels_
print("The Homogeneity is %0.2f" % metrics.homogeneity_score(labels, kmeans.labels_))
print("The Completeness is %0.2f" % metrics.completeness_score(labels, kmeans.labels_))
print("The V-measure is %0.2f" % metrics.v_measure_score(labels, kmeans.labels_))

#Step 11: Further analysis into each cluster (e.g. to find which cluster is worthy of special treatment)
print (reversed_data_transf.groupby('Cluster').agg({
    'Not being able to stop or control worrying':['mean','min','max'],
    'Little interest or pleasure in doing things':['mean','min','max'],
    'Feeling down, depressed, or hopeless':['mean','min','max']}))
```

Fig. 4.48 Codes for performing k-means clustering in Python

increases, the WCSS value decreases. We would select the K value based on this WCSS decrease rate. Let me make it more clear for you. In the graph in Fig. 4.49, from clusters 1 to 2 to 3, you will find a huge fall in WCSS. After three, the drop is relatively slower, and hence we chose 3 to be the optimal value for K.

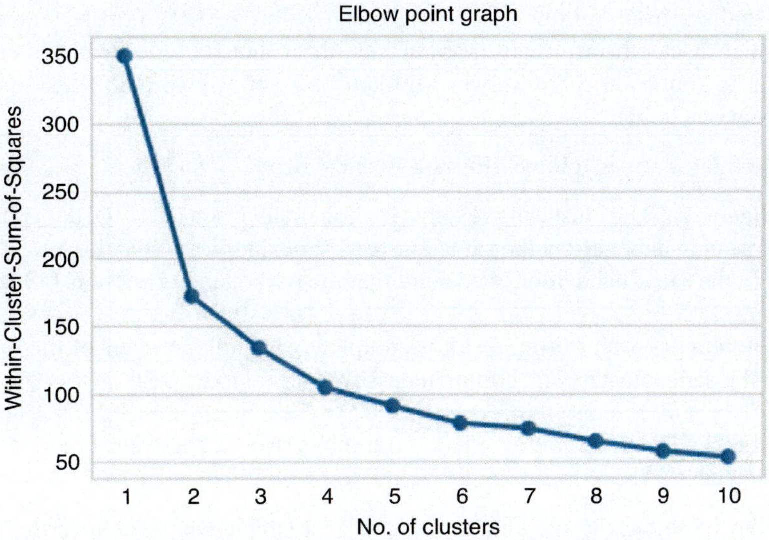

Fig. 4.49 Elbow point graph

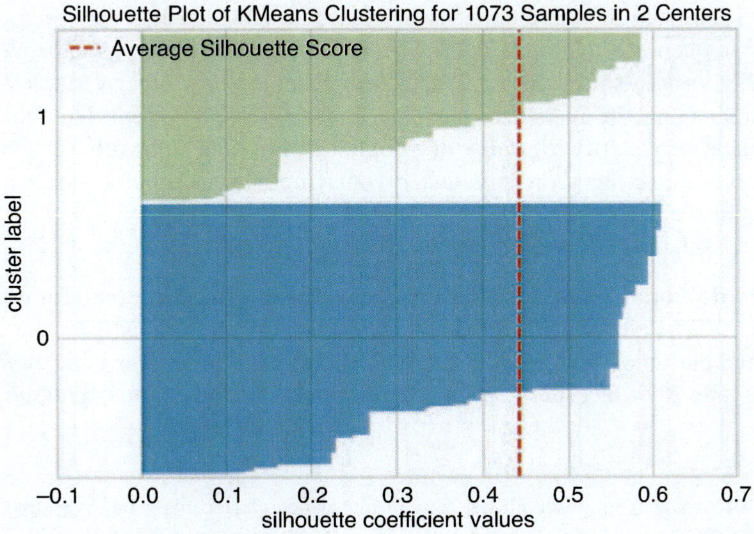

Fig. 4.50 A plot showing silhouette scores

2. **The silhouette method:** Computes the density of clusters estimated by the k-means model. Produces a coefficient score between 1 and −1 (where 1 indicates highly dense clusters). If the silhouette score is −1, then it indicates a completely incorrect clustering. In our case, we found a score of 0.46 (see Fig. 4.50). The take-home message is that: the closer our silhouette coefficient is to 1, the better it is (as denser the clusters are). We used the Silhouette

Visualizer from Yellowbrick (Bengfort & Bilbro, 2019). For the more curious reader, in-depth information about how the silhouette coefficient is calculated can be found on their website (https://www.scikit-yb.org/en/latest/api/cluster/silhouette.html).

The figures are as follows (Figs. 4.49 and 4.50):

Homogeneity The clustering result is homogenous when all of the data points with the same class label are in the same cluster – for example, all the donors in the dataset with the same behaviour. We found that the homogeneity score is 0.85.

Completeness A clustering result is complete when all members of the same class are in the same cluster. The completeness was found to be 0.26.

V-measure The harmonic average of the above two (i.e. between homogeneity and completeness). The V-measure was found to be 0.39.

Table 4.4 shows the average score on each variable under investigation. It informs us of the nature of each cluster. Suppose you notice that all the variables are present in all the clusters. But the proportion (as indicated by the mean or average of that variable) is variable. So, in case you are interested in studying anxiety-related symptoms. Then from the table above, it is clear that cluster 1 has the members with a higher frequency of *Not being able to stop or control worrying*, and hence you will probably wish to study cluster 1 further by checking how other variables in cluster 1 are higher than the average of those in cluster 0. Table 4.4 seems like the clustering was made between two groups of people: healthy and unhealthy because if you compare, the average scores of cluster 0 on symptoms are much lower than the ones under cluster 1.

The resultant scatter plot looks as shown in Fig. 4.51.

Some Additional Thoughts For the convenience of learning (and for demonstrating the 3-D scatter plot), I have included only three variables from the dataset to form the clusters. The variables are "Not being able to stop or control worrying", "Little interest or pleasure in doing things" and "Feeling down, depressed, or hope-

Table 4.4 Analysis into each cluster (e.g. to find which cluster might be interesting for further investigation)

	Not being able to stop or control worrying			Little interest or pleasure in doing things			Feeling down, depressed or hopeless		
Cluster	Mean	Min	Max	Mean	Min	Max	Mean	Min	Max
0	0.9	0	3	0.65	0	2	0.77	0	3
1	2.3	0	3	2.1	0	3	2.23	0	3

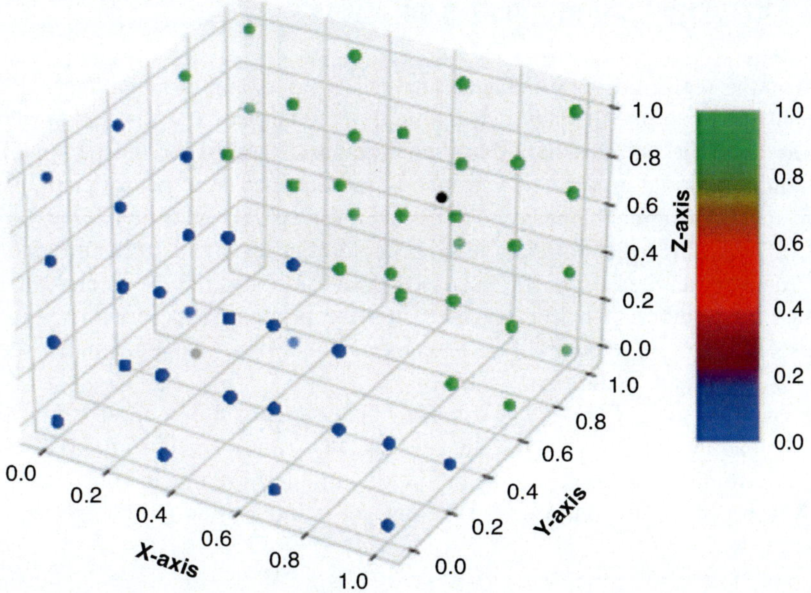

Fig. 4.51 Scatter plot for the two clusters

less". If you want, you could include more (potentially relevant) variables, and the k-means algorithm can handle all of that.

Bonus

Diagnostic manual for mental disorders (e.g. DSM-**5**) suffer from diagnostic heterogeneity. This means that patients diagnosed with the same disorder might report various symptoms. Likewise, patients diagnosed with different disorders might report similar kinds of symptoms. In simpler words, the boundaries of each disorder often fail to group patients into homogenous categories. Maybe you are interested in studying the extent of diagnostic heterogeneity in psychiatric diagnoses. So, maybe you interviewed or surveyed and collected data on all the symptoms experienced by **500** people diagnosed with major depressive disorder and/or post-traumatic stress disorder and/or bulimia nervosa. Then you can use an elbow point graph to determine the number of clusters in a dataset. Suppose the diagnostic manuals would have worked well. In that case, you should get the elbow of the curve at **3**, which would have meant there are three clusters (i.e. major depressive disorder and/or post-traumatic stress disorder and/or bulimia nervosa). Still, more often, you will find it's not the case. Given the studies on diagnostic heterogeneity, you might find two clusters using the elbow graph or any other method. After that, you can run the k-means to estimate the k-means (setting K = **2**) and estimate the homogeneity, completeness and V-measure to test the clustering quality. If I were to guess, I would say that these parameters will be low or moderate at their best.

4.5 Principal Component Analysis (PCA)

PCA is the preferred type of factor analysis for personality researchers. PCA has been used to examine the factor structure of a standardised (norm) dataset of healthy American subjects. This has identified five basic factors for the personality that explain an optimal amount of variance in the original data: neuroticism, extraversion, openness, agreeableness and conscientiousness (as reviewed in Goekoop et al., 2012). We will learn about PCA in this chapter. Earlier in one of the previous chapters (data pre-processing), we discussed dimensionality reduction and how it can reduce the number of predictors, reducing the problems that come with a sparse dataset (curse of dimensionality). We can do PCA for such dimensionality reduction too. The codes are as follows:

From the code in Fig. 4.52, we can say that there are simple three steps. The first step is loading the dataset. Note that we did PCA on normalised data. The second step is assigning the target and the predictor variables. Finally, the third step is using PCA to transform the variables in the dataset into the principal component. If you

In [14]:
```
#Step 1: Load the data
import pandas as pd
data = pd.read_csv("C:\\Users\\ghosh\\Downloads\\Normalised_Dataset_Tutorial_ML.csv", header=0, index_col=0)
data.head()
```

Out[14]:

	Age	Little interest or pleasure in doing things	Feeling down, depressed or hopeless	Feeling nervous, anxious or on edge	Not being able to stop or control worrying	Dummy_Young	Dummy_Middle-aged	Dummy_Elderly	Suicide_Attempt
0	0.371429	1.000000	1.000000	1.000000	1.000000	1	1	0	1
1	0.885714	0.000000	0.000000	0.000000	0.000000	0	0	1	0
2	0.428571	0.333333	0.333333	0.333333	0.666667	1	1	0	0
3	0.314286	0.301587	1.000000	1.000000	1.000000	1	0	0	1
4	0.400000	0.333333	0.333333	0.666667	0.449275	0	1	0	1

In [15]:
```
#Step 2: Assign the target (dependent) variable and the independent variables
feature_names = ["Little interest or pleasure in doing things", "Feeling down, depressed or hopeless", "Feeling nervous, anxious
X = data[feature_names]     #predictors
y = data["Suicide_Attempt"] #predicted
```

In [16]:
```
#Step 3: Using PCA to transform the variables in the dataset to principal components
from sklearn.decomposition import PCA
pca = PCA(n_components=2)
NewComponents = pca.fit_transform(X) #Doing PCA on X
TransformedData = pd.DataFrame(data = NewComponents, columns = ['principal component A', 'principal component B'])
TransformedData.head()
```

Out[16]:

	principal component A	principal component B
0	1.244072	-0.027247
1	-0.753455	-0.108243
2	0.084739	-0.150064
3	0.902447	0.177677
4	0.128968	0.189195

In [17]:
```
#Step 4: Add the target variable back
finalData = pd.concat([TransformedData, data[['Suicide_Attempt']]], axis = 1)
finalData.head()
```

Out[17]:

	principal component A	principal component B	Suicide_Attempt
0	1.244072	-0.027247	1
1	-0.753455	-0.108243	0
2	0.084739	-0.150064	0
3	0.902447	0.177677	1
4	0.128968	0.189195	1

Fig. 4.52 Codes for performing principal component analysis (PCA)

wish, you can add the target variable as shown in step 4. Then it will be possible for you to implement a machine learning model in the next step to predict the target from these principal components (new predictors) if that's what you seek.

4.6 Rule Mining

A study attempted to investigate help-seeking behaviour, that is, the characteristics of students who are open to seeking counselling and those who are not, in a sample of secondary school students in Singapore. The researchers hoped to find relationships between the students' attitudes, opinions and behaviours and whether they sought or would seek counselling help or not. They used association rule mining (ARM) for this exploratory study because it allowed them to examine combinations of variables that characterise students in relation to the research question. Specifically, the goal was to discover frequently occurring attitudes, opinions and behaviours that describe help-seeking behaviour. The study found several interesting results. For example, problems such as schoolwork and loneliness surfaced for those open to seeing a counsellor. More information about this study can be found in Goh and Ang (2007). In addition, I have personally used Apriori algorithm for association rule mining to explore the relationship of auditory hallucinations with other symptoms. Hopefully, these give you an idea of where you can use association rule mining.

The Apriori is an algorithm for frequent itemset mining and association rule learning over relational databases. Traditionally, it was frequently used in the

```
In [9]: #Import libraries, packages, and modules
        import pandas as pd
        import numpy as np
        from mlxtend.frequent_patterns import apriori
        from mlxtend.frequent_patterns import association_rules
        from mlxtend.frequent_patterns import apriori, fpmax, fpgrowth
        import matplotlib.pyplot as plt
        import seaborn as sn

In [10]: #Load dataset
         dataset = pd.read_csv('D:\\NAYA_symptoms111.csv', header = [0])
         dataset.set_index('ID', inplace=True) #specify the index column

In [12]: #To convert all the positive datavalues into one and all the datavalues which are zeros or are in negative as zeros.
         def encoding (xyz):
             if xyz <= 0:
                 return 0
             if xyz >= 1:
                 return 1

         rule_basket = dataset.applymap(encode_units)

In [13]: #Generate frequent item sets that have a support of at least 1% to filter out the most important sets
         frequent_itemsets = apriori(rule_basket, min_support=0.01, use_colnames=True, verbose=1)

...

In [14]: #Generate the rules with their corresponding support, confidence, lift, leverage and conviction.
         rules = association_rules(frequent_itemsets, metric="lift", min_threshold=1)
         rules.head(100)

...

In [15]: rules.to_excel('C:\\Users\\ghosh\\Downloads\\results.xlsx')
```

Fig. 4.53 Codes for association rule mining

marketing literature and practice, such as market basket analysis, to predict which consumer items people are frequently likely to buy together. However, in the context of the mental health literature, it has been infrequently used, particularly on survey data (Goh & Ang, 2007; H.N & Dr. N.V., 2019; Zhao et al., 2017).

The code from my study was somewhat as shown in Fig. 4.53.

The Apriori function extracted frequent itemsets for association rule mining to analyse data and answer the questions. Apriori is a popular algorithm in commerce used to extract frequent itemsets in association rule learning. The Apriori algorithm is designed to operate on a database containing transactions, such as purchases by store customers. But we demonstrate that it can be as invaluable to analyse the strength of relationships between symptoms in qualitative data (e.g. textual narratives) as correlation is to quantitative data (e.g. survey) as mentioned above.

Rule mining generates many rules, and not all of them are important. Therefore, special attention has been given to selecting association rules and variables in association rule modelling. The Apriori algorithm allows us to use selection criteria that reflect the potential utility of association rules. In other words, different techniques within rule mining can generate potential rules and then can be manually reviewed to filter out the most important ones. There is no clear "best" statistic – instead, statistics should be chosen based on individual datasets and applications (Wright et al., 2010).

If you run the above code, you will get the result as follows (Fig. 4.54):

This is the result I found in one of our studies. The metrics indicate the following (refer to Table 4.5 for a quick summary):

- **Support** (Agrawal et al., 1993): Support is used to measure the abundance of frequency (often interpreted as significance or importance) of an item set in a database. We refer to an itemset as a "frequent itemset" if your support exceeds a specified minimum-support threshold. We specified that only associations that achieve a certain minimum level of support of 0.01 are included in the mining model. This was to ensure a highly meaningful result. It is also one of how we

```
In [20]: #Generate the rules with their corresponding support, confidence, lift, leverage and conviction.
         rules = association_rules(frequent_itemsets, metric="lift", min_threshold=1)
         rules.head(100)
Out[20]:
```

	antecedents	consequents	antecedent support	consequent support	support	confidence	lift	leverage	conviction
0	(fear_)	(cry_)	0.125858	0.045767	0.011442	0.090909	1.986364	0.005682	1.049657
1	(cry_)	(fear_)	0.045767	0.125858	0.011442	0.250000	1.986364	0.005682	1.165523
2	(eating-issues_)	(fear_)	0.046911	0.125858	0.010297	0.219512	1.744124	0.004393	1.119994
3	(fear_)	(eating-issues_)	0.125858	0.046911	0.010297	0.081818	1.744124	0.004393	1.038018
4	(afraid)	(fear_)	0.028604	0.125858	0.028604	1.000000	7.945455	0.025004	inf
...	...	...	...	...	...	...	...	...	...
63	(depression)	(schizophrenia)	0.100686	0.066362	0.010297	0.102273	1.541144	0.003616	1.040002
64	(anxiety)	(depression)	0.067506	0.100686	0.024027	0.355932	3.535054	0.017231	1.396303
65	(depression)	(anxiety)	0.100686	0.067506	0.024027	0.238636	3.535054	0.017231	1.224769
66	(depression)	(paranoia)	0.100686	0.069794	0.011442	0.113636	1.628167	0.004414	1.049463
67	(paranoia)	(depression)	0.069794	0.100686	0.011442	0.163934	1.628167	0.004414	1.075649

68 rows × 9 columns

Fig. 4.54 Result of association rule mining

Table 4.5 Explanation of the commonly used metrics in association rule mining

Metrics	Sample question
Support	How frequently did a symptom X-symptom Y occurred as pair among all the relationships?
Confidence	What is the likelihood of symptoms X and Y happening together?
Lift	How likely symptom Y is reported when symptom X is reported while controlling for how popular symptom Y is?
Conviction	To what extent the consequent is highly dependent on the antecedent? If items are independent, the conviction is 1.

controlled the number of associations that were created. The result was that 402 associations were generated. We argue that support should not be used for identifying the important associations because association rules may score highly on support simply because the symptoms are very frequent in the database, even if the relationship between X and Y is weak. The standard practice is that the researcher arbitrarily sets the support (and confidence) criteria arbitrarily to filter out and focus the discussion on the "important" rules. For example, suppose the support threshold is set to 0.1 (10%). In that case, the frequent itemset is defined as a group of items that appear simultaneously in at least 10% of all associations in the database.

- **Leverage** (Piatetsky-Shapiro, 1991): Leverage computes the difference between the observed frequency of A and C appearing together and the frequency that would be expected if A and C were independent. A leverage value of 0 indicates independence. In the filtered dataset, 228 out of 402 associations had leverage = 0 and therefore were removed, leaving 174 associations for the final analysis. Therefore, support and leverage were mainly used to filter out the most interesting associations.
- **Confidence** (Agrawal et al., 1993): The confidence of a rule A->B is the probability of seeing the consequent in a transaction, given that it also contains the antecedent. Suppose the confidence is 1 in the database. In that case, it means that for 100% of the associations containing auditory hallucination and symptom X, the rule is correct (100% of the times a patient reports symptom X, auditory hallucination is reported as well). From our results, three out of 174 associations have confidence of 1. Likewise, a confidence score of 0.42 indicates 42% of the times when a narrative reported symptom X. They also reported auditory hallucination. Therefore, confidence is a measure of reliability and can be interpreted as a conditional probability estimate.
- **Lift** (Brin et al., 1997a)**:** It measures the "interestingness" of an association. The value of lift is that it considers both the support of the rule and the overall dataset. A lift ratio larger than 1 implies that the relationship between the antecedent and the consequent is more significant than expected if the two sets were independent. The larger the lift ratio, the more significant the association. We analysed the 174 associations using the lift. These are the main metrics used in the current study, with confidence and conviction metrics as supporting. When two symptoms are independent of each other, no rule can be drawn involving those two symptoms. If some rule had a lift of 1, it would imply that the antecedent's prob-

ability of occurrence and consequent is independent of each other. There were no such associations in this filtered list.

- **Conviction** (Brin et al., 1997b): A high conviction value means that the consequent is highly dependent on the antecedent. For instance, in the case of a perfect confidence score, the denominator becomes 0 (due to $1 - 1$), for which the conviction score is defined as "inf". Similar to lift, if items are independent, the conviction is 1. In the filtered dataset, none of the associations had a conviction score of 1. However, three of the associations had a conviction of infinity (for those three associations which had a perfect confidence score of 1). This was used as a supporting metric to the lift in the above code. Together, you can use the lift to discuss the results with conviction metrics as an adjunct metric. Such metrics enable quantification of the strength of association between two variables/symptoms from the current data and thereby are expected to generate the rules/associations that are of interest to the psychopathology literature.

Note, in this lesson, the terms antecedent and consequent have been used to keep at par with the marketing literature from where the current rule mining concepts come. However, these antecedents and consequents do not have a causal meaning. For example, during market basket analysis, it might be inferred that "bread and butter are bought together" – this indicates these two products were billed on the same day by the same customer(s). In this case, one of them might be assigned as an antecedent (e.g. bread) and the other as consequent (e.g. butter). However, this does not indicate causality but co-occurrence. No attempt is made to infer which one is picked first, "bread" or "butter". In fact, some experts use LHS and RHS instead of these terms.

Ethics and Legality

The infamous incident of Facebook-Cambridge Analytica exemplifies the ethics and legality side of using data science by psychologists. In case you don't know what happened, Cambridge Analytica was contracted to the Trump campaign for the US Presidential Election. What Cambridge Analytica did was segment Facebook users using psychographics.

Psychologist Aleksandr Kogan had once collected 270,000 personality tests completed by Facebook users through an online app he had created for research purposes. He sold Cambridge Analytica access to it. Kogan's data also collected Facebook data from the test-takers' friends, which summed up to a total of 50m people in the dataset. So, many people got their data in the dataset even without taking the test. Another academic, Michal Kosinski, got involved. He is a computational social scientist and assistant professor in organisational behaviour at the Stanford Graduate School of Business. He predicted someone's personality profile from LIKEs on Facebook. Based on these and other sources of data, Cambridge Analytica built personality profiles for more than 100m registered US voters and developed ad variations on different political themes tailored to different personality profiles. Then they used it for targeted advertising which supposedly had a role to play in Trump's victory. Regardless of the controversy or fate of Cambridge Analytica, behavioural/psychological analytics will stay and so will target marketing on the Internet. But do you see the power of psychologists in influencing a country's ruler? As you must have heard, "In this world, with great power, there must also come great responsibility" (Lee & Ditko, 1962). So, just because you can, doesn't mean you should.

For this reason, I thought of explicitly mentioning this here. In this book, I taught you some of the potential tools and techniques that can be used on peoples' data, which will generate a massive impact on their lives. Make good use of this power, and make a positive impact on the lives of people.

Legality

At times we do not collect data directly from people. Instead, we gather data from the websites about numbers, such as traffic to a website or data from its users. This is called web scraping. Consider the example of a research study on sexual fantasy by extracting approximately 250,000 text-based erotic fantasies from a user-generated website (Seehuus et al., 2019). Using such website data is great because it prevents social desirability bias. But remember that the website is a property of a person or an organisation. Picking up data from someone's property can make it trespass and theft. It's safer just to write an email and ask the owner if you can use their data.

Legal issues might arise when you violate someone's copyright or scrap the data from a website that does not allow you to. To play safe, I recommend contacting the website owner and getting written permission on whether you can scrap their contents for your research work. If they say NO, then do not proceed. Instead, check for other websites. If you want to read more on web scraping, I recommend you read Mancosu and Vegetti (2020). Likewise, if someone else collected a dataset, wrote an article, shot a photo or recorded a video, it is their property. They own the copyright. Unless they make it open to all, they hold the rights to the content. So, just because it is available online in the public domain does not make it free to use by anyone. Again, to avoid infringing someone's copyright, make sure you ask the person and get their written consent.

Ethics

Just because it is not illegal does not make it ethical. The British Psychological Society (2017) published the Ethics Guidelines for Internet-mediated Research. It speaks about respect for individuals and communities' autonomy, privacy and dignity, scientific integrity, social responsibility, maximising benefits and minimising harm. Ensure that your research study follows such ethical guidelines or at least similar ones (your country might have a different guideline). As researchers, we are accountable and responsible for what happens to our research participants and the world at large because of our research publications. Of course, there will be times of moral or ethical dilemma. In such cases, seek advice from multiple senior colleagues and the ethical board of the institution, and then modify your strategy accordingly. The end goal is to ensure that our research does better than harm this world and its people. That's why we are researching in the first place. Do you disagree?

To end, machine learning is a powerful tool for analysing data. But the entire process, from collecting the raw data to reporting the inferences, involves many legal and ethical risks. For example, what if you grab data from a website which does not permit you to do so in their *terms of service*, or what if the way you reported the inferences enables a reader to track down one of the individuals in your dataset? If done without being mindful of these factors, a simple, innocent choice in the process can result in a lawsuit. In the court of law, *Ignorance is not innocence but sin* – so this lesson introduces some of the possible ethical and legal issues you need to be mindful of while working on a data science project. That said, you still need to consult with your lawyer or speak with your line manager to gain a comprehensive knowledge of all the other possibilities and what you need to do to prevent them. If you are based in the university, then usually, these are taken care of along the process by clearing the requirements of the ethical committee and by attending training programmes available in your country. In the United Kingdom, the Safe Researcher Training might be helpful to learn more about such concerns with data science projects and how to prevent mishaps. With that knowledge, close this book (read the conclusion if you want to), start typing some codes on your computer and use your datasets to experiment. A wide range of mysteries and research questions are awaiting you to be unfolded.

References

Agrawal, R., Imielinski, T., & Swami, A. (1993). Mining association rules between sets of items in large databases. In Proceedings of the AGM SIGMOD Conference on Management of Data (pp. 207–216). Washington, DC.

Armour, C., McGlinchey, E., Butter, S., McAloney-Kocaman, K., & McPherson, K. (2020). The COVID-19 psychological wellbeing study: Understanding the longitudinal psychosocial impact of the COVID-19 pandemic in the UK; A methodological overview paper. *Journal of Psychopathology and Behavioral Assessment, 43*(1), 174–190. https://doi.org/10.1007/s10862-020-09841-4

Barnett-Itzhaki, Z., Elbaz, M., Butterman, R., Amar, D., Amitay, M., & Racowsky, C. et al. (2020). Machine learning vs. classic statistics for the prediction of IVF outcomes. *Journal of Assisted Reproduction And Genetics, 37*(10), 2405–2412. https://doi.org/10.1007/s10815-020-01908-1

Bengfort et al., (2019). Yellowbrick: Visualizing the Scikit-Learn Model Selection Process. *Journal of Open Source Software, 4*(35), 1075, https://doi.org/10.21105/joss.01075

Bixler, R., & D'Mello, S. (2015). Automatic gaze-based user-independent detection of mind wandering during computerized reading. *User Modeling And User-Adapted Interaction, 26*(1), 33–68. https://doi.org/10.1007/s11257-015-9167-1

Bleidorn, W., & Hopwood, C. J. (2019). Using machine learning to advance personality assessment and theory. *Personality and Social Psychology Review, 23*(2), 190–203.

Brin, S., Motwani, R., & Silverstein, C. (1997a). Beyond market baskets: generalizing association rules to correlations. In SIGMOD '97: Proceedings of the 1997 ACM SIGMOD international conference on Management of data (pp. 265–276). Retrieved 21 August 2022, from https://doi.org/10.1145/253260.253327

Brin, S., Motwani, R., Ullman, J., & Tsur, S. (1997b). Dynamic itemset counting and implication rules for market basket data. In SIGMOD '97: Proceedings of the 1997 ACM SIGMOD international conference on Management of data (pp. 255–264). Retrieved 21 August 2022, from https://doi.org/10.1145/253262.253325

British Psychological Society. (2017). *Ethics guidelines for internet-mediated research.* Retrieved February 26, 2021, from https://www.bps.org.uk/sites/www.bps.org.uk/files/Policy/Policy%20-%20Files/Ethics%20Guidelines%20for%20Internet-mediated%20Research%20(2017).pdf

Bulling, A., & Zander, T. O. (2014). Cognition-aware computing. *IEEE Pervasive Computing*. 13, 80–83. https://doi.org/10.1109/MPRV.2014.42

Bzdok, D., Altman, N., & Krzywinski, M. (2018). Statistics versus machine learning. *Nature Methods, 15*(4), 233–234. https://doi.org/10.1038/nmeth.4642

Dwyer, D., Falkai, P., & Koutsouleris, N. (2018). Machine learning approaches for clinical psychology and psychiatry. *Annual Review of Clinical Psychology, 14*(1), 91–118. https://doi.org/10.1146/annurev-clinpsy-032816-045037

Farmer, A., McGuffin, P., & Spitznagel, E. (1983). Heterogeneity in schizophrenia: A cluster-analytic approach. *Psychiatry Research, 8*(1), 1–12. https://doi.org/10.1016/0165-1781(83)90132-4

Goekoop, R., Goekoop, J., & Scholte, H. (2012). The network structure of human personality according to the NEO-PI-R: Matching network community structure to factor structure. *Plos ONE, 7*(12), e51558. https://doi.org/10.1371/journal.pone.0051558

Goh, D., & Ang, R. (2007). An introduction to association rule mining: An application in counseling and help-seeking behavior of adolescents. *Behavior Research Methods, 39*(2), 259–266. https://doi.org/10.3758/bf03193156

Hassan, L. M., Shiu, E., & Shaw, D. (2016). Who says there is an intention–behaviour gap? Assessing the empirical evidence of an intention–behaviour gap in ethical consumption. *Journal of Business Ethics, 136*, 219–236.

Henderson, J. M., Shinkareva, S. V., Wang, J., Luke, S. G., & Olejarczyk, J. (2013). Predicting cognitive state from eye movements. *PLoS ONE* 8:e64937. https://doi.org/10.1371/journal.pone.0064937

Hoppe, S., Loetscher, T., Morey, S., & Bulling, A. (2018). Eye movements during everyday behavior predict personality traits. *Frontiers in Human Neuroscience, 12*. https://doi.org/10.3389/fnhum.2018.00105

Jacobucci, R., & Grimm, K. (2020). Machine learning and psychological research: The unexplored effect of measurement. *Perspectives on Psychological Science, 15*(3), 809–816. https://doi.org/10.1177/1745691620902467

Jordan, M. I., & Mitchell, T. M. (2015). Machine learning: Trends, perspectives, and prospects. *Science, 349*(6245), 255–260.

Kapoor, S., & Narayanan, A. (2021). (Ir)Reproducible Machine Learning: A Case Study. Reproducible.cs.princeton.edu. Retrieved 21 August 2022, from https://reproducible.cs.princeton.edu/irreproducibility-paper.pdf

Koul, A., Becchio, C., & Cavallo, A. (2018). PredPsych: A toolbox for predictive machine learning-based approach in experimental psychology research. *Behavior Research Methods, 50*, 1–16.

Kroenke, K., & Spitzer, R. (2001). The PHQ-9: A New Depression Diagnostic and Severity Measure. *Psychiatric Annals, 32*(9), 509–515. https://doi.org/10.3928/0048-5713-20020901-06

Lee, S., & Ditko, S. (1962). Spider-Man. *Amazing Fantasy, 15*, 13.

Li, J., & Tong, X. (2020). Statistical hypothesis testing versus machine learning binary classification: Distinctions and guidelines. *Patterns, 1*(7), 100115. https://doi.org/10.1016/j.patter.2020.100115

Liu, N., & Salinas, J. (2017). Machine Learning for Predicting Outcomes in Trauma. *Shock, 48*(5), 504–510. https://doi.org/10.1097/shk.0000000000000898

Lundberg, S., & Lee, S. (2017). *A unified approach to interpreting model predictions*. In 31st conference on neural information processing systems. Long Beach, CA, USA.

Mancosu, M., & Vegetti, F. (2020). What you can scrape and what is right to scrape: A proposal for a tool to collect public facebook data. *Social Media + Society, 6*(3), 1–11. https://doi.org/10.1177/2056305120940703

Medvedeva, M., Vols, M., & Wieling, M. (2019). Using machine learning to predict decisions of the European Court of Human Rights. *Artificial Intelligence and Law, 28*(2), 237–266. https://doi.org/10.1007/s10506-019-09255-y

Meehan, A., Latham, R., Arseneault, L., Stahl, D., Fisher, H., & Danese, A. (2020). Developing an individualized risk calculator for psychopathology among young people victimized during childhood: A population-representative cohort study. *Journal of Affective Disorders, 262*, 90–98. https://doi.org/10.1016/j.jad.2019.10.034

Meyer, A., Zverinski, D., Pfahringer, B., Kempfert, J., Kuehne, T., Sãndermann, S., et al. (2018). Machine learning for real-time prediction of complications in critical care: A retrospective study. *The Lancet Respiratory Medicine, 6*(12), 905–914. https://doi.org/10.1016/s2213-2600(18)30300-x

Nimon, K. (2012). Statistical assumptions of substantive analyses across the general linear model: A mini-review. *Frontiers in Psychology, 3*. https://doi.org/10.3389/fpsyg.2012.00322

Orrù, G., Monaro, M., Conversano, C., Gemignani, A., & Sartori, G. (2020). Machine learning in psychometrics and psychological research. *Frontiers in Psychology, 10*. https://doi.org/10.3389/fpsyg.2019.02970

Piatetsky-Shapiro, G. (1991) Discovery, Analysis, and Presentation of Strong Rules. Knowledge Discovery in Databases. AAAI/MIT Press, Cambridge, 248, 255–264.

Priya, A., Garg, S., & Tigga, N. (2020). Predicting anxiety, depression and stress in modern life using machine learning algorithms. *Procedia Computer Science, 167*, 1258–1267. https://doi.org/10.1016/j.procs.2020.03.442

Richter, T., Fishbain, B., Markus, A., Richter-Levin, G., & Okon-Singer, H. (2020). Using machine learning-based analysis for behavioral differentiation between anxiety and depression. *Scientific Reports, 10*(1). https://doi.org/10.1038/s41598-020-72289-9

Rimfeld, K., Kovas, Y., Dale, P. S., & Plomin, R. (2016). True grit and genetics: Predicting academic achievement from personality. *Journal of Personality and Social Psychology, 111*, 780–789.

Rosenbusch, H., Soldner, F., Evans, A., & Zeelenberg, M. (2021). Supervised machine learning methods in psychology: A practical introduction with annotated R code. *Social and Personality Psychology Compass, 15*(2). https://doi.org/10.1111/spc3.12579

Roy, A., Nikolitch, K., McGinn, R., Jinah, S., Klement, W., & Kaminsky, Z. (2020). A machine learning approach predicts future risk to suicidal ideation from social media data. *Npj Digital Medicine, 3*(1). https://doi.org/10.1038/s41746-020-0287-6

Rozek, D., Andres, W., Smith, N., Leifker, F., Arne, K., Jennings, G., et al. (2020). Using machine learning to predict suicide attempts in military personnel. *Psychiatry Research, 294*, 113515. https://doi.org/10.1016/j.psychres.2020.113515

Sato, J., Moll, J., Green, S., Deakin, J., Thomaz, C., & Zahn, R. (2015). Machine learning algorithm accurately detects fMRI signature of vulnerability to major depression. *Psychiatry Research: Neuroimaging, 233*(2), 289–291. https://doi.org/10.1016/j.pscychresns.2015.07.001

Seehuus, M., Stanton, A., & Handy, A. (2019). On the content of real-world sexual fantasy: Results from an analysis of 250,000+ anonymous text-based erotic fantasies. *Archives of Sexual Behavior, 48*(3), 725–737. https://doi.org/10.1007/s10508-018-1334-0

Shi, T., Seligson, D., Belldegrun, A., Palotie, A., & Horvath, S. (2004). Tumor classification by tissue microarray profiling: Random forest clustering applied to renal cell carcinoma. *Modern Pathology, 18*(4), 547–557. https://doi.org/10.1038/modpathol.3800322

Spitzer, R., Kroenke, K., Williams, J., & Löwe, B. (2006). A Brief Measure for Assessing Generalized Anxiety Disorder. *Archives Of Internal Medicine, 166*(10), 1092. https://doi.org/10.1001/archinte.166.10.1092

Tasmim, S., Dada, O., Wang, K., Bani-Fatemi, A., Strauss, J., Adanty, C., et al. (2020). Early-life stressful events and suicide attempt in schizophrenia: Machine learning models. *Schizophrenia Research, 218*, 329–331. https://doi.org/10.1016/j.schres.2019.11.061

The UK National Health Service (NHS). (2021). *A guide to good practice for digital and data-driven health technologies*. Retrieved February 19, 2021, from https://www.gov.uk/government/publications/code-of-conduct-for-data-driven-health-and-care-technology/initial-code-of-conduct-for-data-driven-health-and-care-technology

Wright, A., Chen, E., & Maloney, F. (2010). An automated technique for identifying associations between medications, laboratory results and problems. *Journal Of Biomedical Informatics, 43*(6), 891–901. https://doi.org/10.1016/j.jbi.2010.09.009

Zhao, L., Xu, T., Hao, F., & Dong, X. (2017). Positive and negative association rules mining for mental health analysis of college students. *EURASIA Journal Of Mathematics, Science And Technology Education, 13*(8). https://doi.org/10.12973/eurasia.2017.01011a

Chapter 5
End Note

There are different approaches to analysing data analysis. Statistics and machine learning are two examples. None is superior than the other. The choice of the kind of data analysis to perform is guided by what kind of research question you are asking or what type of dataset you have. Traditionally, statistics had been used in psychological sciences, typically to analyse survey data. Until recently, many quantitative psychological researchers relied solely on surveys. But today, we have so much more to analyse. Thanks to the recent technological progress. For example, every time we shop, socialise, play and work online, our actions and behaviours are getting recorded somewhere. Most often, people are unaware of this because they don't read the websites' and browser companies' terms of service and privacy policies. Other times, we willingly give our private information to organisations, such as the banks, to access their services (such as a mortgage). Such data are often collected, stored and shared by and within the organisations for both commercial and non-commercial purposes (e.g. operational convenience). For example, social media platforms might give such data to marketing agencies which would then display targeted advertisements to potential customers. Other times, healthcare centres (e.g. hospitals) or government agencies use similar data to offer you and the society a better and safer service. For a psychologist, this means that they have now got access to data about millions of people in different domains: hospitals, schools, businesses, charities and so forth. They just need to contact the relevant organisation and seek their permission to use the dataset for academic purposes. If they allow, there is a ready dataset of thousands of people sitting on your computer waiting to be analysed. With these new and diverse data sources, relying only on statistics is restrictive to the kind of investigations that can be performed. But after reading this book, now that you know machine learning, the variety of research questions and data sources you can use have increased as well. However, you can read a book on "100 ways to swim", but you won't learn swimming unless you actually dive into the water. So, I cannot emphasise enough how much you need to engage yourself in analysing data. Therefore, the next step from here would probably be to try finding

C. Ghosh, *Data Analysis with Machine Learning for Psychologists*,
https://doi.org/10.1007/978-3-031-14634-3_5

different kinds of data, utilising the knowledge learnt in this book and trying to solve different research questions. I am confident that you will experience problems. There will be errors, and at times they will be frustrating. But the good part is you can always refer to the Internet and look up for a potential solution. There might be times when you will be stuck in a problem for weeks. I had been there myself several times. Such experiences developed my patience, and to be honest, I found it somewhat addictive – so don't give up after a few bumps on the road. Trust me. It would be a worthwhile learning experience.

Congratulations on making it to the end. To make a quick summary, we learnt about Python programming, data pre-processing and machine learning for data analysis.

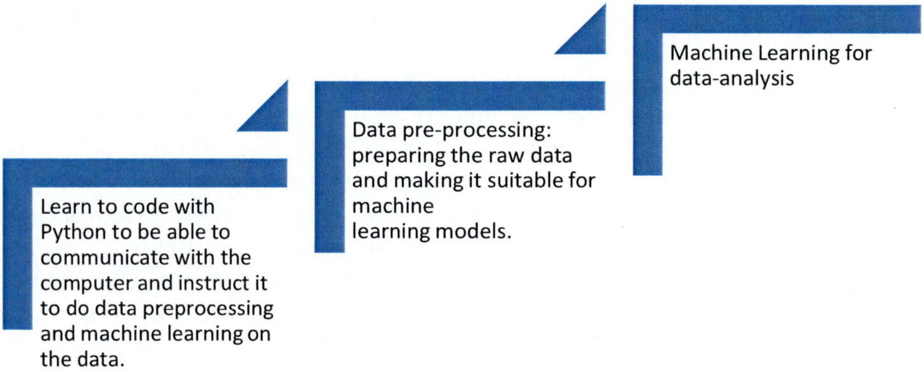

We also got a quick overview of research design and statistics and learnt about ethics and legal issues. I am more than happy to welcome you to email me and let me know your learning journey. If you are stuck somewhere in the book or feel something hard to understand or if something does not make sense in this book, please do email me. I will try my best to clarify as soon as I can. That said, you should practise the habit of searching for your answers on Q&A websites such as https://stackoverflow.com/. It will make you more self-reliant and resolve your doubts quicker.

You might also have some requests or questions outside the scope of this book. For example, you are probably looking for some help with machine learning for analysing your dataset. In that case, I provide professional consultancy services to academics, businesses and students. If you need any help or wish to collaborate, the best way to get in touch is to email me at ghoshchandril@gmail.com. Even if you are a student - don't feel shy. Get in touch.

Compared to where you started when you first opened the book, by now, you should have arrived at a position where you know how to implement the machine learning codes for your research. You might experience the **Impostor syndrome** but believe in yourself. Now you know more than what you knew when you just started this book. You will be surprised to find yourself as one of those fewest psychologists with this special knowledge. Chances are you are the only one in your department because this subject is still new in our discipline, and you are one of the first. Take some pride in that. Teach your colleagues and students whatever little you have learnt because the more you teach, the more you know. Don't hesitate to test your new superpower. This is only the beginning.

Contact the Author for Help

I am fairly easy to reach out. If you have any questions related to the content of this book, or just want to say a "HI!", you can email me at ghoshchandril@gmail.com or get in touch with me over the following channels:

- Facebook: https://www.facebook.com/DrChandril
- LinkedIn: https://www.linkedin.com/in/chandril-ghosh/
- Twitter: https://twitter.com/Chandril7

The biggest way in which you can help me – if you feel inclined – although you certainly don't have the obligation to do so – is by leaving a kind review on the website you bought this book from – or just by spreading the word. I would love more psychology students and researchers to use machine learning in their research so that they don't have to screw up their research project as I did years back – wish you all the best – with much love.

Index

A
Association rule mining (ARM), 149–151

B
Basics, 6–9, 26, 29, 41, 42, 47, 76, 88, 116, 136, 148

C
Classifiers, 29, 61, 87, 89, 92, 99, 107–116, 119, 121–123, 131, 133
Clustering, 1, 2, 13, 15, 17, 88, 89, 92, 94–96, 99, 100, 104, 136–148
Coding, 6, 7, 33, 35, 47, 52, 54
Conclusion, 3, 88, 126, 154
Contact author, 153, 157, 159

D
Data cleaning, 58–66, 84
Data pre-processing, 2, 56, 57, 66, 67, 102, 105, 132, 148, 158
Data reduction, 61, 62, 74–84
Data transformation, 57

H
Help, 7, 14, 15, 20–23, 26, 27, 42, 50–53, 57, 59, 64, 83, 87, 89, 90, 93, 94, 98, 101, 102, 119, 123, 127, 130, 132, 133, 136, 139, 140, 149, 158, 159

L
Learning to code, 7

M
Machine learning, 1–11, 13–15, 17, 23–26, 29, 30, 49–51, 55, 58, 62, 66, 71–73, 77, 79, 83, 87–117, 119, 120, 123, 125–133, 135, 136, 149, 157–159

P
Principal component analysis (PCA), 83, 84, 96, 97, 136, 148–149
Python 3, 5, 34–36
Python programming, 2, 4, 52, 158

R
Regressor, 61, 87, 89, 92, 99, 107, 116, 123, 129–131, 133, 134

S
Summary, 12, 57, 67, 150, 158
Supervised, 11, 83, 87, 89, 90, 92–95, 97, 99–112, 125, 136
Support, 6, 110–112, 132, 150, 151

U
Unsupervised, 11, 17, 83, 87–90, 92–95, 97, 99, 100, 104, 136

C. Ghosh, *Data Analysis with Machine Learning for Psychologists*,
https://doi.org/10.1007/978-3-031-14634-3